DESIGNING DIGITAL PRODUCTS FOR KIDS

DELIVER USER EXPERIENCES THAT DELIGHT KIDS, PARENTS, AND TEACHERS

Rubens Cantuni

Apress®

Designing Digital Products for Kids

Rubens Cantuni
Monza, Italy

ISBN-13 (pbk): 978-1-4842-6289-4   ISBN-13 (electronic): 978-1-4842-6287-0
https://doi.org/10.1007/978-1-4842-6287-0

Managing Director, Apress Media LLC: Welmoed Spahr
Acquisitions Editor: Shiva Ramachandran
Development Editor: Rita Fernando
Coordinating Editor: Nancy Chen

Cover designed by eStudioCalamar

Distributed to the book trade worldwide by Springer Science+Business Media New York, 1 New York Plaza, New York, NY 100043. Phone 1-800-SPRINGER, fax (201) 348-4505, e-mail orders-ny@springer-sbm.com, or visit www.springeronline.com. Apress Media, LLC is a California LLC and the sole member (owner) is Springer Science + Business Media Finance Inc (SSBM Finance Inc). SSBM Finance Inc is a **Delaware** corporation.

For information on translations, please e-mail booktranslations@springernature.com; for reprint, paperback, or audio rights, please e-mail bookpermissions@springernature.com.

Apress titles may be purchased in bulk for academic, corporate, or promotional use. eBook versions and licenses are also available for most titles. For more information, reference our Print and eBook Bulk Sales web page at http://www.apress.com/bulk-sales.

Any source code or other supplementary material referenced by the author in this book is available to readers on GitHub via the book's product page, located at www.apress.com/9781484262894. For more detailed information, please visit http://www.apress.com/source-code.

Printed on acid-free paper

To Yui, who fills my life with joy and laughter.

Contents

About the Author

Rubens Cantuni is an Italian digital product designer with 15 years of experience across two continents. He has won an Emmy Award in the "Outstanding Interactive" category, a Webby Honoree, and several Parents' Choice awards and Teachers' Choice awards for his work on digital products for children. His experience spans from agencies to startups to big corporations, covering multiple design roles for a wide variety of clients in different industries. He also writes about design on Medium and Builtin.com and has past experience as a character designer and illustrator, freelancing for many companies worldwide.

Acknowledgments

This book wouldn't be possible without the help of many people.

The ones who agreed to be bothered by me and my questions for the "Industry Insights" interviews. Some of them are friends and former colleagues, others heard from me for the first time because of this book, nonetheless, despite being incredibly busy professionals, have been so gracious to answer all my questions, providing incredible value to this project.

My daughter Yui and my wife Kanako, who tolerated my being busy during the long realization of this book, with my skipping breakfast to write before work, my writing time during weekends, my "sorry, no Netflix tonight, I gotta finish a chapter."

I wish to thank all the nonfiction authors in the design field, who inspired and motivated me to start this project and made me believe it was possible.

Lastly, a special thanks to Shivangi Ramachandran, Rita Fernando, Nancy Chen, and Apress for believing in a first-time author and his project and giving him this fantastic and educational opportunity.

Introduction

It was a late warm, sunny afternoon in April 2017, in Los Angeles. For the first time in my life I rented a tuxedo, and I was getting ready for an event that required a black-tie outfit. I was about to leave my apartment in Venice, headed to the Pasadena Civic Auditorium for the Daytime Emmy Awards, where I was nominated in Outstanding Interactive, a category that, until a few weeks before that day, I wasn't even aware it existed.

How did an Italian digital product designer end up at an event where fellow nominees included the likes of Guillermo Del Toro? It was all thanks to a digital product for kids I worked on.

That night, the team working on the StoryBots' digital products, which included me, went home with a heavy golden statue. But my experience with these kinds of projects started a few years before that, when I designed the first educational apps for a startup, or websites like the one that Swatch did for their kids' collection, among many others. Why then did I tell you the story about the Emmys, though? To brag? Well, yes, a bit of that (it won't happen again in this book), but to also let you know that digital products for kids are not a niche, a minor genre. You can get very rewarding results from them; maybe not a mansion in Malibu, but as a designer you get to create products that educate and entertain little humans, the future of our species, all the while perhaps accumulating some awards and recognitions along the way. Ain't that something?

In 2018, I came back to Italy and I wrote a series of articles on the same topic of this book on Medium. All the interest they generated, the requests for advice and consulting, made me realize that there is a need for a reference book on this subject. When I was taking my first baby steps (pun intended) in designing digital products for kids, there had been a lot of trial and error, a lot of assumptions, and a lot of questions without answers. I write this book, wishing I had this at that time.

How This Book Is Organized

The concept is to follow an ideal process for the genesis of a digital product aimed toward children. Starting with research, so knowing the industry and learning about your target; then moving on to concept and design, from UX and interaction to UI, animation, and sounds; user testing and validating your idea, before releasing it; evaluating possible business models and marketing solutions. In the end we'll also look at interesting technologies and concepts that go beyond the screen.

Each chapter works independently and you can jump directly to the one you're most interested in. You can move back and forth and easily use this book as a reference whenever you need information on a particular phase of product development.

What We Won't Be Talking About

The primary audience for this book are digital product designers,[1] and, with them in mind, I defined the table of contents for this book. We'll talk about some technicalities in Chapter 6, when evaluating the right platform for our digital product, but we won't be talking about coding. For two reasons:

I don't think product designers should code. I think they should have an awareness of how code works, but they are designers and that's what they should stick to.

I'm not a developer (yes, I made some mobile games, but that's a story for another book), so I'm not qualified to talk about that.

Both excellent reasons to skip the subject and leave it to other books (there are plenty out there, each one with a more mysterious name than the next one).

We won't be talking about video games (and trust me, I love them and I would talk about them for hours—ask my wife), unless specifically targeted to K-12 children. What I mean is: Can a 10-year-old play *Minecraft*? Absolutely. Can a 6-year-old play *Candy Crush*? Yes, of course. Were they made specifically for kids? No. Then we won't talk about them in this book.

Would a college student play with *Peppa Pig: Holiday*? Unless the student is a reader of this book and wants to design an app for kids, my guess is no, they wouldn't. So we'll probably talk about that game here.

You got the idea. To be 100% honest, I'll mention some not-for-kids video game titles here and there, when they help to explain an idea, but we'll never analyze them under the lens of products for children.

---

[1]Cantuni, Rubens. "What's a Digital Product Designer?" *Medium*. UX Collective, 18 Jan. 2020. Web. 24 May 2020. <https://uxdesign.cc/whats-a-digital-product-designer-50e09a125efb>.

Will This Book Be Still Useful in the Next Year? And the Year After That? And 5 Years from Now?

I can't guarantee your little nephew will find a job, years from now, thanks to this book. But, besides some information strictly connected to current technologies (things like devices and development platforms), most of the concepts in this book have a very long expiration date and some of them are even timeless—unless kids of the future will be born with superpowers or technology will be so advanced it'll be able to read our mind.

Is Designing for Children Really That Different from Designing for Adults?

It is. The framework is certainly similar, virtually the same, but there are particularities in how to handle tests, in what the best practices for UX and UI are, and so on.

Designing for kids depends on many things, including factoring in the rapid pace of physical and mental development in children. While adults have more or less the same motor and cognition skills that stay the same for many years, kids develop quickly and just a few months could be enough for considerable advances in their proficiencies. This has a huge impact on how we design the user experience for children.

When using an app, adults are more interested in a task-oriented approach, meaning they know what they want to accomplish and they want to do it fast and easy. Children, on the other hand, prefer an experience-oriented approach; they are in it for the journey, not much for the result.

How Much About Child Development and Education Do I Need to Know in Order to Design Digital Products for Them?

The more you know the better, but no one expects a designer to be also a pediatrician or an infant psychologist or a pedagogist. You should have a basic understanding of how each age group corresponds to different stages of cognitive and physical development and what these involve. If you are building an educational experience (but remember that it's not mandatory for it to be strictly educational), you'll need the support of experts in the field, during the conceptual phase and then during testing.

Do I Need to Be a Parent to Design for Kids?

Not at all. In fact, I designed several award-winning products for kids before my daughter was even born. For sure it's been very fulfilling, when she was finally old enough to try them out, to find out that she adored them.

Being a parent is (besides one of the most amazing and difficult experiences in life) helpful because it gives you the opportunity to empathize more with children, to have a daily taste of how they think, what they need, how their little minds see the world. These are things you can also see during tests, to some extent, but a "firsthand" experience as a parent can represent a plus. But again, the support of experts, the testing sessions, the interviews with parents can suffice. Do not consider designing a digital product for kids as a good enough reason to have babies!

Sometimes, being a parent could even be counterproductive. The risk is to make assumptions and base decisions on what we observe on one or two (or ten in some parts of the world) children and think these are valid for all of them. Children are people, each one is different, and the observation of our own kids can't validate anything.

Why Design Apps for Kids?

Find the Right Motivation

If you are reading this book, you probably already have your own reasons as to why you want to design apps for kids. Maybe you already work in the field and want to hear another point of view or have a broader vision on topics outside of your day-to-day. Perhaps you are considering starting a business in this industry and you want to know the basics, or, why not, you're a parent that just wants to make informed choices about the products your child uses.

But let me share with you three of my reasons that might serve you as additional inspiration and motivation. If you have the same motives, well, hey! You are not alone! If you have different ones, good! Now you have more.

Business Opportunities

> A pessimist sees the difficulty in every opportunity;
>
> an optimist sees the opportunity in every difficulty.
>
> —Winston Churchill, Former Prime Minister of the UK

© Rubens Cantuni 2020
R. Cantuni, *Designing Digital Products for Kids*,
https://doi.org/10.1007/978-1-4842-6287-0_1

This might not sound as the most noble of reasons, but let's not forget that we're talking about products and design here, not art. Design ultimately needs to sell, products need to be sold, so business opportunities should be one of the primary reasons to undertake the creation of an app or website of this kind. It's important to get to know the market, the competitors, and the target audience and see if there are opportunities for our product to get its place and make a profit.

Now let's say we're not interested in making money and we just want to make the world a better place. We have an idea and we just want to give it away for free to a nonprofit organization or just by ourselves. If by our analysis it looks like this idea wouldn't have a market, it being a paid product or not doesn't make much difference. Even if it's free, having no market means there is no actual need for it. If there's no need for it, if it doesn't solve any problem, if it doesn't serve any need, there must be something fundamentally wrong in our concept, or maybe we picked the wrong target audience.

So money or not, it's important that our idea is marketable, otherwise it's a piece of art, it's a statement with no actual purpose.

There are many opportunities for kids' products. Björn Jeffery made an estimate[1] of the total market size for kids' apps, resulting in around $80 billion globally, in 2019. That's a pretty nice amount of money.

On October 30, 2019, the streaming colossus Spotify announced a kid-friendly version of its app, specifically targeting kids, with a different UX/UI and curated content. The same goes with YouTube that launched YouTube Kids on February 15, 2015. These two examples tell us a couple of interesting things:

- If the big guys of tech are interested in providing experiences for the youngest ones, then most probably there is a market for that. For them it might just represent a way to breed their audience for their primary products or to please parents and serve the entire family strengthening the brand perception. But they wouldn't bother if there was no request for such products.

- Not all products for children need to be educational. Of course, songs or video can have educational purposes, but we can consider these as apps for entertainment.

Speaking of educational products though, here are some interesting numbers: according to a research by HolonIQ[2] education is a $6T industry growing to $8T by 2025. This is about education as an industry, not referred only to digital products and for kids specifically, but those are some staggering numbers.

---

[1]Jeffery, Björn. "The Kids App Market, Part 1: A Strategic Overview." *Björn Jeffery.* N.p., 31 May 2019. Web. 24 May 2020. https://www.bjornjeffery.com/2019/05/31/the-kids-app-market-a-strategic-overview/.
[2]Holoniq. "Global Education in 10 Charts." https://www.holoniq.com/. N.p., 2019. Web. 2020.

Talking specifically about educational apps for kids, I give you two examples:

- Fifty percent of US classrooms[3] (in 2018) use an educational app from a startup called Kahoot.

- ABCmouse, the leading e-learning platform for kids, is (as of mid-2020) #39 of the top-grossing apps in the United States. And I'm talking about the overall top-grossing chart of the entire App Store catalogue, including streaming services, video games, and so on. That's a very remarkable result.

Even though making a splash on the App Store or Google Play Store is much more difficult now than years ago, the kids' segment can still offer chances that others can't (e.g., the gaming category is way more overcrowded and led by giant companies able to invest tons of money in marketing and promotion).

One of the biggest and most common problems with these products is the lack of quality and real educational value, when marketed as such. For this reason, there is a lot of room for improvement on this front, and quality could represent the advantage against a competition of suboptimal alternatives.

Educational Value

It is a miracle that curiosity survives formal education.

—Albert Einstein, theoretical physicist

In the previous section, I mentioned how the lack of standards for kids' products, marketed as educational, is one of the biggest problems for this industry.

It's right in this lack of regulation that I see the second opportunity. As designers we have the duty to solve people's problems with sensible solutions, even though we don't swear like doctors and law enforcement officers, we

---

[3]Lunden, Ingrid. "Education Quiz App Kahoot Says It's Now Used by 50% of All US K-12 Students, 70M Users Overall." *TechCrunch*. TechCrunch, 18 Jan. 2018. Web. 24 May 2020. https://techcrunch.com/2018/01/18/education-quiz-app-kahoot-says-its-now-used-in-50-of-all-us-classrooms-70m-users-overall/.

should always remember that design should help humans and facilitate progress. While the previous reason for developing apps for kids seemed venal, here I'd like to explain how we can redeem ourselves. Too often these products hide themselves behind being "educational," while in reality there is little to no substance to support such a statement. Parents are way more keen to spend money on apps that are supposed to be good for their children, and what's better than education?

That's why the promise of "learning with fun" is tempting, and digital products for kids abuse this idea just to sell. Another common keyword used by self-proclaimed educational apps is "guilt-free"; this leverages on one of the most common feelings among parents. A 2019 survey by Mashable and What to Expect[4] on 1,700 moms showed how guilt over screen time is one enormous concern.

There is a lot of room for improvement on the educational value for these products, and I feel that designers should now step up their game and really try to think about the well-being of their little audience, before the marketing opportunities that this idea of "educational product" carries.

Over 80,000 apps (surely more at the time you read this) are labeled as educational; very few are backed up by researches that show this quality.

The more we conduct research on our products, working along with pediatricians, educators, researchers, the more best practices will emerge and with them the whole industry will improve on quality, real tested educational value and we'll be able to serve our purpose as designers.

Some advantages of using computers and devices with kids are as follows:

- **Training for the future**: As adults, we know way too well how screens are now ubiquitous to almost any profession as well as being present in many of our hobbies and free-time activities. Keeping in mind the already mentioned concerns about screen time, we can't deny that using devices in schools is a way to develop skills that will help children in their future relationship with technology.

- **Engagement in learning**: This is valid only when there is real educational value in products. Some kids enjoy the interaction aspect of apps and website when learning; it helps keeping them engaged and focus on the activity. The gamification and the idea of getting achievements and unlock content add motivation and stimulate curiosity.

---

[4]Hazlett, Alex. "New Parents Are Anxious about Screen Time, but It Gets Better." *Mashable.* Mashable, 17 Sept. 2019. Web. 24 May 2020. https://mashable.com/article/kids-screen-time-guilt-survey/.

- **Customized and individual learning**: In classroom activities, children are normally subjected to the same methodology regardless of their individual needs. But some kids may benefit from more visual activities, others from auditory experiences, others from reading, and so on. Not to mention children with disabilities or learning difficulties, who may struggle to engage in activities in groups for fear of judgement. Digital products on individual devices can cater for these different needs with a tailored experience.

- **Better performance**: Researcher and data scientist Maya Lopuch found in her research[5] that kids who had included educational apps in their curriculum saw improved performances in Common Core Standards[6] domains. Another study by University of Southern California professor Michelle Riconscente found that by just playing 20 minutes with a math app for 5 days, 5th graders improved their results by 15%, compared with a control group. Another study[7] by Houghton Mifflin found that 20% more middle schoolers scored "Proficient" or "Advanced" in understanding algebra by using an app rather than a textbook. Pretty encouraging results!

These are just some examples of how designers working on products for kids can really get a sense of fulfillment from helping future generations' development. This sense of working to do something good is not that common in today's tech companies, where privacy scandals, undisclosed data mining, and basically any user exploitation happen daily.

It was one of the primary reasons why, in 2015, my wife and I made the decision to pack our lives into four pieces of luggage to move from Italy to Los Angeles' silicon beach (a.k.a. Venice) to join StoryBots and help them develop their digital products for preschoolers.

---

[5]Lopuch, Maya. "The Effects of Educational Apps on Student Achievement and Engagement." (2013). Print.

[6]Council of Chief State School Officers. "Preparing America's Students for Success." *Common Core State Standards Initiative*. 2020. Web. 24 May 2020. http://www.corestandards.org/.

[7]Bonnington, Christina. "IPad a Solid Education Tool, Study Reports." *CNN*. Cable News Network, 23 Jan. 2012. Web. 24 May 2020. https://edition.cnn.com/2012/01/23/tech/innovation/ipad-solid-education-tool.

A Design Challenge

There is nothing like a challenge to bring out the best in man.

—Sean Connery, actor

We discussed how digital products for children desperately need excellent design and why that is important. A very common misconception is that apps for kids are easier than the ones aimed to adults, but this is a very superficial evaluation. Kids' products are often more complicated than others because of the many constraints in terms of user experience, interaction, user interface, and general regulations (by laws and by App Store standards) they are subjected to.

We'll have the chance of analyzing deeply each of those in later chapters; here I'd like just to highlight how these challenges represent an opportunity for designers.

The role of product designer is one that implies a continuous process of learning. In one article[8] I wrote some time ago, I explained how one cannot be a product designer without the hunger for continuous studying and improving. Technology is constantly evolving, at a very high speed, with it fresh opportunities arise and our life is constantly changing. Designing products for kids, being them educational or not, offers lots of challenges and occasions to learn new skills. Making a product simple enough to be used by preschoolers, who can't read, who have a limited dexterity compared to an able adult, can teach us a lot on how to make usable and inclusive experiences for grown-ups as well.

Industry Insight: Interview with Björn Jeffery

Björn Jeffery is the cofounder of Toca Boca, a Swedish app development studio, leader in digital products for kids. They have been the first to introduce the idea of "digital toy." Björn has been also board member for Sago Mini and Rovio, before becoming a full-time independent advisor specialized in kids' media, mobile strategy, and consumer insight.

---

[8]Cantuni, Rubens. "You Can't Be a Product Designer If You Don't Like Studying." *Medium.* UX Collective, 21 June 2018. Web. 24 May 2020. https://uxdesign.cc/you-cant-be-a-product-designer-if-you-don-t-like-studying-fe6fa65fc2d6.

Rubens: When I decided to write this book, the first example of a company with good products for children that came to my mind was Toca Boca. I've been a fan of their apps even before becoming a dad. So, my first question is: What are, in your opinion, the main reasons for Toca Boca's success? What did set it apart from competitors?

Björn: I think it was a combination of factors that made Toca Boca successful—it's hard to look at any single thing and say that it was the sole reason. But to name a few:

Our strategy of making digital toys as opposed to games was quite new when we launched in 2011. It was a completely different way of thinking about apps for kids, and it created a different user experience with less stress and no emphasis on competition. This helped us stand out a lot initially.

We spent a lot of time developing our brand and our graphical identity. There's a specific look and feel to Toca Boca products that are distinctly different from most other kids' products. This was very intentional and contained some components that were very radical in their context—like designing everything in a gender-neutral way and for a diverse audience.

Arguably, the most important factor was our focus on making a delightful experience for kids first. That meant prioritizing their needs in all testing and ensuring the core interactions were fantastic and that they worked as intended in the settings that the kids would use them in. This may sound obvious, but a quick glance across a lot of kids' media will reveal that this isn't something you can assume to be getting. Many products might look good, but they don't actually work from a kid's point of view. The app should work as the kids expect it to and enable superpowers that they didn't know they had. This takes care, attention to detail, and a respect for the craft that is making apps for kids.

R: Before becoming a leader in the industry, how many failed attempts have there been for Toca Boca? What has been the biggest or most frequent mistake in such failures?

B: We failed numerous times in all manner of ways—products, marketing, organization, business models, and more. But that is an inevitable part of running a business. It's a process that you need to go through. The only thing you can hope for is to not repeat the same mistakes several times.

If I were to single out a specific failure, it would likely be our video product Toca TV. I could probably write a book on that process alone. But in brief, it was an example of where we saw the emerging trend of the importance of kids' video early, but we misread the ecosystem and what position we could take within it. The product that we produced simply wasn't good enough given the competition that we were up against. In hindsight it was correct to make a big bet on video, but the product we chose to do that with was not right.

R: Toca Boca has probably been the first company to introduce the concept of "digital toys" (which I love). When we think of apps to play with, our mind immediately goes to video games. Video games, from the most casual to AAA ones, always pose a challenge to the players. There is something they need to do to go further into the experience that, if done wrong, puts an end to it. In digital toys, the experience itself is the goal, it's a sandbox approach that we always had with physical toys, but was quite new in the digital world. Where did the idea come from?

B: Neither Emil Ovemar—my cofounder—nor I had a gaming background, other than that we had played a lot of video games. We came from the web side of things with more of an emphasis on interaction as opposed to strict gaming mechanics. That aspect played in as we could look at the emerging app landscape more broadly than just games. We looked at anything that was using a touchscreen in an interesting manner, and thought about how this could be adapted to kids. Some of it was also simply observational. Emil saw how his own kids used their devices and that they treated it like a toy—often as a part of a different playing experience. LEGO figures going to the cinema, for instance. It sparked the idea that perhaps it is adults thinking that this is a completely new and unique experience but that for kids it may be much simpler. It just needs to be fun! Just like a toy is.

R: Talking about the industry, how do you think it has evolved since the time you first approached it? Do you think digital products for kids today are better than a decade ago? If so, what do you think contributed the most to such evolution?

B: It has obviously grown and professionalized a lot since we started. There's a much stronger focus on monetization and streamlining of the types of concepts that reach a broad audience. I think the overall art and technology quality is significantly better, but the space as a whole is arguably less interesting. There's less art and experimentation going on, or at least it is less visible. The kids' app space could—somewhat ironically—have a lot to gain from being more playful and fun.

R: In this book, one of the first challenges I present to the readers is the complexity of the target audience. We have children, in different age groups, where a 2-year span means the world; we have the parents and then the educators. Such audiences have very different needs and expectations. What is the best way to approach this challenge? How can we find the right compromise?

B: It is a very complicated situation since there are several areas where kids and parents have opposite incentives and wishes for what they want from a product. I think the easiest way is to clearly decide who you are designing for and then prioritize their needs first. Is this a supplemental education product that parents buy for their kids to use after school? Fine—then you need to design it in a way so that parents understand that. The kids preferably need to like it too, but you are choosing the parents first in that scenario. Toca Boca was the opposite to that. We always chose kids first, much to the detriment of many parents that frequently asked if we could make math apps instead. But when kids gave us requests, they were never about math. So that's what steered our decision.

R: Monetizing products for kids might be a complicated matter. Even though there are specialized "kid-safe" ads networks nowadays, I think that advertising can't be the primary source of revenues for children's products, especially if aimed to younger kids. In-app purchases are gated, requiring the kids to ask the parents to buy digital goods. Then we have the good old onetime payment and subscriptions. The choice here heavily depends on the nature of the app. If the product is constantly updated with new content, a onetime payment can't be the strategy. If the product is not updated regularly with new content, a subscription model is not justifiable. In your experience, is there a model that works best? In the case of a onetime paid app, the free "lite" version + paid "premium" version can be a successful strategy? What about bundles?

B: This is a question which can only really be solved on an app level. It's hard to make broad generalizations since it depends on all the factors that you mention in the question. But I will say that the macro has clearly shifted from paying upfront toward creating a subscription. This in turn changes the products as well as they need to justify the subscription costs. That's why we are currently seeing so many "World" apps that are essentially bundles of a lot of different activities in one. A smaller developer, however, might not have the resources to initially invest enough in order to compete in such a market. So for them, a free + in-app purchase model would likely be better. It is a very complex question to answer unfortunately.

R: Alright, last question. What do you think the future of digital products for kids will be? We recently saw big tech companies showing more interest in younger users, for example, Spotify with their app for kids, or Google that launched a new section for kids on the Play Store with "teacher approved" apps. Will this trend continue? Will we see more "kids' versions" of products originally made for adults?

B: I think there's a broader acknowledgment that kids use the Internet, whether you designed it for them or not. Privacy and regulatory concerns also drive this development a lot. I wish I could say I thought this was coming from vision and interest, but I think it is likely more an issue of compliance.

The most interesting products for kids in the future will still likely come from small teams experimenting without a lot of preconceived notions around what would work and not. Designing products with care, love, and attention to detail. There's a lot left to be done.

Chapter Recap

- Designing digital products for children is a good business opportunity. The market is crowded, but growing and hungry for good products. The business of education is also growing and rich of opportunities.

- Many apps are labeled as "educational," but just a minority has real value. Creating an educational product for kids is an opportunity to put our design skills up for a good cause.

- Children's products are challenging to design. They offer opportunities for learning, improving, and becoming better, more inclusive designers.

Before You Start, Know the Industry

Understand What's on the Market and How It Works.

In this chapter, we'll look at the landscape of apps for kids. We'll try to keep a broad view to get a grip on the general basic knowledge you should have before even starting to ideate a product.

Why is this important? Well, it's always essential to know the market for a product we're going to develop, but in this case I believe it's exceptionally significant because, while as adults we constantly use many products, for productivity, entertainment, communication, wellness, and more, we are not accustomed to using products for kids in our daily tasks. Therefore, while we already have general knowledge on how a music streaming app works, because we know it from a user perspective, we know little about video streaming

© Rubens Cantuni 2020
R. Cantuni, *Designing Digital Products for Kids*,
https://doi.org/10.1007/978-1-4842-6287-0_2

apps for kids, because we are not the target for such apps. Clearly, for any kind of digital product, there are many more things to know besides the experience as a user, but not having hands-on practice adds another layer of complexity. Parents might have an advantage here, but developing a more critical eye for this category of digital products requires some effort and analytical skills, because, again, they are not made with grown-ups in mind.

We'll also look at the differences between educational and entertainment, and we'll examine various genres of products for kids and their peculiarities. Lastly, I'll share with you some tips on where to look for ideas and inspiration for your products.

Entertainment, Educational, Edutainment

Everybody loved 'The A-Team' because it was entertainment, pure and simple.

—Mr. T, actor

Apps can fall into many categories. It just takes a quick glance at the App Store or Google Play Store to find apps for productivity, social networking, streaming music and videos, messaging, fitness, gaming, news, travel, cooking, and many others. These apps are for grown-ups and categorizing them is an efficient way to make sense of the incredible amount of products out there. Apps for kids are one of these categories, but they often blend in between gaming and education.

Many games, when they respect a series of rules and limitations (we'll discover which ones in a later chapter), can be categorized as kids' products, even though they were created with adults in mind.

One of the biggest issues we still face today, with the App Store or Google Play Store, is that apps for children are all under the same umbrella. Sometimes there are collections like "ideal for preschoolers" or "apps for kids 3 to 5" and so on, but by looking inside you can see the situation is absolutely chaotic. Apps to teach spelling are together with video apps and games and who knows what else. Sure, these are all products for kids, but they are all different in purpose!

Can you imagine looking for a weather app and then going on the App Store of your choice to just find a giant "apps for adults" category, and everything is inside that, Spotify, Airbnb, Netflix, Microsoft Excel, Google Maps... How crazy would that be? Yes, the number of apps for children is way smaller than the amount of apps for adults, but it's still a pretty big number.

Even though apps for children are a category (or sometimes more of a tag) by itself, we can identify subcategories that the stores seem to ignore. There's a fragmentation pertaining to the age group, and then we can define another fragmentation about the purpose of the app. We'll talk about age groups later, while in this section we'll have a look at the purpose a kids' app can have.

There are two major groups your app could fall into—*entertainment* and *educational*—and a third one which is a combination of the two, called *edutainment* (a portmanteau of **edu**cational and enter**tainment**).

Gaming, music, videos, dancing, karaoke, and all this kind of things have, as their prime purpose, to entertain the user. Counting, spelling, reading, coding (yes, coding), and so on are educational.

The topic of quality in educational apps (that we briefly touched upon in Chapter 1) is very important, because there are consequences to false claims and deregulation—the main two being

1. Not providing a good experience, or even a detrimental one to children, in a worst-case scenario.

2. It can harm the confidence of parents in buying these products. If they have no tools to understand which app is really educational, which app was really meant to help their child's development, they'll eventually end up downloading something that doesn't keep up to what it promises; this can cast a bad light over the entire market of apps for children and parents will lose trust in it altogether.

The third group, *edutainment*, is the most sought-after solution for children's products. It's probably every educator's (and parent's) dream to teach to children by maintaining a high level of interest without lapsing into boredom. Trying to add gaming and fun elements is not a new strategy, it's been done for years, in schools and in products for children before the digital age, for example, in board games. You don't need to be a parent or a caregiver to understand a preschooler might not find counting very exciting without adding some fun to it.

In fact, gamification for the purpose of learning is very common even in apps aimed to adults. Look at *Duolingo*, for example, features like badges and achievements are there to add a sense of challenge and reward when completing a series of activities. We actually use this technique in apps unrelated to education, think, for example, at the health app on the Apple Watch. Every day you have to close three rings, and that alone, the progression of the rings completing the circle is a powerful incentive, and when you do for several days in a row, you get a special badge as a reward. We will focus more on gamification in Chapter 5.

Even apps that are leaning on the entertainment side of the scale, for example, YouTube Kids, push educational content at the forefront. Why? Because, as I mentioned in Chapter 1, education sells and helps to deal with screen time concerns.

Edutainment is not necessarily a perfect formula of 50% educational value and 50% entertainment. The mixture can vary from product to product in different proportions (see Figure 2-1).

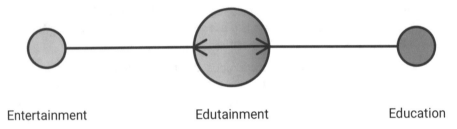

Entertainment   Edutainment   Education

Figure 2-1. *The educational component can vary. A product can lean more toward the fun or vice versa*

It has to be a fine balance, learning and having fun are often (for most kids) on two opposite poles, and you don't want to push too much on the learning side and making a product boring, because a boring product doesn't engage kids, and if kids don't want to use it, they waste parents' money, and angry parents leave nasty reviews—trust me. But you also don't want to create just a funny, silly experience, with no value for the kid's development and not marketable to parents. There are many factors that influence how a product is perceived as educational: how you market it, if it's endorsed by experts and/or studies, if it gets used in schools, if it's under a brand or intellectual property with credibility in educational products and content, for example, *Sesame Street*.

Regardless of the concept you might have in mind, it's ideal to mix these two powerful ingredients in the right amount.

Types of App Experiences

Provided that, as mentioned in the previous section, the best cocktail for a successful app is a combination of education and fun, let's look at what sort of experiences are out there for kids. This doesn't want to be an extensive list, you are free and encouraged to create new products that won't fit in any of the following, but if we'd have to create folders on our tablet with the apps on the market right now, I think we could create four big groups and probably make a very good percentage of them fit into one of these.

So let's take a look at them and for each we'll identify the key traits.

Educational

Straight educational apps (even including the gamification layer) are some of the most appreciated by parents and educators and as a result also among the profitable ones.

I won't reiterate on the importance of quality for these apps (quality is obviously important for any app though); I assume the previous sections did the job and convinced you that real educational value is important and should be one of the things we pursue in our design. For this reason, educational apps can be challenging. Designers are rarely also educators, so it helps to enlist the help of a subject matter expert, such as a certified teacher, to help when designing an educational app; a competent professionals that can help us shape the experience, the activities and their difficulty level, according to the target age group and according to standards we might want to respect, they have to be believable and authoritative. To do so, they usually require investments to get credibility. Certifications or awards like the Teachers' Choice Award can do the job (this could give an edge when marketing our product—more on this later).

One of the projects I worked for was called *StoryBots Classroom*. StoryBots is an intellectual property (IP) created by *JibJab* brothers Gregg and Evan Spiridellis (now acquired by Netflix) where funny little creatures live inside electronic devices and their life purpose is to answer kids' questions. Around this IP, we created a series of products, including a TV show, multiple digital and physical books, an iPad app, a website, and an e-learning platform for schools, and I was involved in these last three projects.

While the app and the website were more focused on the entertainment side of the IP, the platform for kindergartens, called *StoryBots Classroom*, was heavily focused on education. When I started my job there, I already had experiences in products for kids (some specifically made for young kids, others were mobile games also suitable for kids, but for adults as well), but never on products for schools with such a heavy focus on the educational side. Plus the whole curriculum of activities had to comply with US government guidelines for preschool education, called Common Core Standards, and for me, not being an educator and not even American, this was a total mystery at the beginning. I started educating myself on the matter, of course, but the real help came from our educator consultant Nina Neulight. I worked closely with her to understand the Common Core Standards requirements, specifically about math, and to learn what sort of activities were indicated for preschoolers. How they can learn to count? Can they do simple additions? Up to what numbers? How can we teach them counting with coins? And many, many others. Nina was my go-to person when I had questions and needed immediate feedback on ideas, but we had literally thousands of real teachers trying out the exercises in a closed beta-testing program; with them, we could conduct quantitative researches through periodic polls and by collecting their direct feedback.

So, educational apps, the real ones, the ones made with the purpose of teaching something and not just reassuring parents about their insecurities, involve the work of people with experience in education, and specifically in education for the target age of the product we're creating.

Simulations

The game of "pretending" is a typical kids' activity; imitating adults is a natural learning technique in kids' development process. Imitation is a way to play, and playing is an activity we have in common with all mammals, and not just them according to studies[1, 2] so having fun is not just a human thing.

Some apps take inspiration by this to make simulation games where kids can experience things like being veterinarians, hairdressers, musicians, doctors, firefighters, or even airport security officers.

Two dominant players in these products are Toca Boca (`https://tocaboca.com/`) and Dr. Panda (`https://drpanda.com/`). These apps are usually structured as "worlds" or "environments" where kids can experience a diverse range of activities, each one made of simple tasks. Some of them can be just endless repetitive tasks (repetition is key in kids learning, they don't get as easily bored as adults when doing the same things over and over), while others can include mini-games with a real objective to achieve.

These games have also a layer of education, but the gaming component is usually higher than products focusing mainly on learning.

What I like about Toca Boca's products is that, in most of them, kids have direct experience on the activity, without the mediation of a character performing the actions. So, for instance, in *Toca Hair Salon*, kids can cut characters' hair directly with a touch of their finger, without the intermediation of an *avatar* doing the actions. It's the same difference as FPS (first-person shooter) like the *Call of Duty* series and third-person action video games, like *Tomb Raider* or *Super Mario Bros.*

The creativity component is also another big feature of these games. In *Toca Tailor*, kids can create outfits for the characters using a series of tools and libraries of patterns, or even snap photos to create their own patterns.

[1]"Is Play Unique to Mammals?" *Popular Science*. Popular Science, 11 June 2009. Web. 25 May 2020. `https://www.popsci.com/scitech/article/2009-06/play-unique-mammals/`.

[2]Zaraska, Marta. "Non-mammals Like to Play, Too." *Discover Magazine*. Discover Magazine, 21 Nov. 2019. Web. 25 May 2020. `https://www.discovermagazine.com/planet-earth/non-mammals-like-to-play-too`.

Some other games can be just explorative experiences, like the *Sago Mini* series, where children can explore a simple and delightfully illustrated world to find funny animations and interactions.

The Dr. Panda series is also very interesting. The library of apps is very extensive, and the activities range from firefighter to ice cream truck to restaurant to farm, candy factory, train conductor, and many more—each of these involves a series of mini-activities related to the job the app is about. Such activities require simple repetitive interactions, but the kid has a general idea of what the job is about, what are the tasks involved, and basically to learn how the world works.

Simulation apps are products that more than other resemble video games as we commonly intend them. But there are some differences though that it's important to notice and that make the difference between a video game and a digital toy. Simulations of this kind are more digital toys (see the next section) than video games.

Sandbox

With "sandbox" we mean all those products that offer an environment in which the users are given tools to freely express their creativity. The real-world version of this are, well, sandboxes of course, but also LEGO and all building blocks, Play-Doh and all kinds of clay, and so on. These games appeal to children and adults, both the real and the virtual ones.

The greatness of these games is the freedom they give to players; some might include mini-games to unlock achievements and new tools, but mostly the gameplay involves a lot of freedom and exploration.

Dr. Randy Kulman, a veteran psychologist specialized in child's therapy, notes[3] "Child psychologists and play therapists have used these techniques for many years in their work with children. A common tool in a play therapist's office is something called a sand tray. Sand trays are small sandboxes with a variety of characters and objects that allow kids to describe the major relationships and events in their lives. Sand trays often have toy buildings and structures to facilitate expression of thoughts and concerns about home and school and are considered being a powerful way to help kids explore and articulate feelings about difficult issues."

---

[3]Kulman, Randy. "Like the Real Thing, Sandbox Games Can Promote Freedom and Creativity: The Power of Play: Toca Boca." *The Power of Play | Toca Boca.* N.p., n.d. Web. 25 May 2020. https://tocaboca.com/magazine/sandbox-games/.

Freedom and creativity are the major drivers of these games' success among children (and adults as well); it can translate in exploring, building, and destroying things. Studies highlight[4] the benefits of free play, a form of play that is not structured by adults, children can make their own rules and establish their relationships with others, improving collaboration as well as problem-solving skills. Being allowed to do the wrong thing and retry is at the base of Montessori method;[5] "Montessori educational practice helps children develop creativity, problem solving, critical thinking, time-management skills, care of the environment and each other. It can prepare them to contribute to society and to become fulfilled persons."

With digital products, Minecraft is maybe the most popular example among kids. There are even books and manuals for children on how to build stuff in *Minecraft* (not to mention YouTubers and Twitch streamers). Differently from *Minecraft, Toca Blocks* is a sandbox game aimed specifically to kids, and even though it might be a little too complicated for a preschooler, its sandbox nature allows it to be appealing for a wide range of ages.

Content Library Apps

What we analyzed so far were active forms of entertainment, but there are very successful apps that provide a passive experience. This means that the child is not actively engaged in the progression of the experience.

YouTube, Netflix, Disney, Cartoon Network, Nickelodeon, and PBS are all major players in video content that offer some kind of app (or dedicated section) for kids. The educational value varies depending on the IP and the quality of the writing. The popularity of a TV series and characters and the content offer are a major driver in the success of such apps.

In this case, the saying that *content is king* is absolutely 100% true. The experience has to be simple and enjoyable; the content has to be easily accessible, well organized, categorized, but the big role here is played by the content itself. But as designers we have to create the right container to make this content shine. There are lots of things to take in considerations even when we are building around a library of content. For example, the media player is a key component of this genre of applications. When I worked on the

---

[4]Lahey, Jessica. "Why Free Play Is the Best Summer School." *The Atlantic.* Atlantic Media Company, 20 Aug. 2018. Web. 25 May 2020. https://www.theatlantic.com/education/archive/2014/06/for-better-school-results-clear-the-schedule-and-let-kids-play/373144/.

[5]"Montessori's Method." *MONTESSORI, The Official International Montessori Site for Theory and Teacher Trainings Information.* N.p., n.d. Web. 25 May 2020. http://www.montessori.edu/.

StoryBots app for iOS, the first parts we designed, and probably the most important ones, were the media player and the library of content. The scope of the media player is to make the content available to the viewers; there is a set of basic functionalities people expect to perform on a player, such as

- Play and stop the content
- Go to the previous or the next piece of content in a playlist
- Scrubbing along the timeline
- Make the content full screen or minimized

In particular, our player had the added complication of being usable for both video content and animated books, and for both kinds of content, part of the library involved the special feature called "Starring You," in which kids could use their face and become the main character in the story. So not all players are the same; sometimes they need to be tailored for the content they are going to show.

The library is usually the other crucial component in such products. The main tasks of a library of content are

- Presenting the content in an organized manner
- Making content accessible through search and categories
- Highlighting the newest pieces of content
- Suggesting content based on users' preferences and history
- Providing the possibility of creating a list of favorites and/ or customized playlists

Products that don't require interaction from the user could induce in what we call "binge-watching." Binge-watching means that provided a continuous stream of content, the user could lose track of time and keep on watching for hours. This is detrimental to health (and we especially want to avoid this for our young audience).

We can find an example of this in the Netflix app, when, if we don't interact with it for a few episodes of a series, a dialog appears on screen asking us if we're still watching. Nintendo consoles, such as the Wii and Switch, took this even further by suggesting, after a lengthy gaming session, to take a break.

So it's a good practice to offer tools for parents, like timers to lock the app after a set amount of time. Some new TV sets and mobile devices offer this functionality natively in their operating system, but a quick and accessible way to provide this feature within our product is surely a good idea. The investment in terms of development is very low and it shows parents that we designers, developers, and brands care about their kids.

Utilities, Wellness, and More

Other apps fall into smaller categories, like wellness, communication, and utilities, and it's hard to define requirements and characteristics, as they could be basically anything. These represent a minority (at least at the moment) of the children's catalogue both on the App Store and Google Play Store. For this reason, I just briefly mention some interesting examples here.

Communication apps are often used by kids and adults together, like the messaging app for families, *Kinzoo*, which, unlike Facebook's *Messenger Kids*, is built thinking about children from the ground up, and it's not a "porting" of an app for adults, meaning it's not a by-product of something that was originally made for adults and was transformed into something suitable for kids. *Kinzoo* puts a great deal of effort in children's privacy and safety and collects only the bare minimum of personal data. Or *ClassDojo*, a communication platform for teachers to communicate with kids and their parents.

Recently, mindfulness apps for children are gaining popularity. *Mindful Powers* is an award-winning app specifically created for children, just like *Stop, Breathe & Think Kids* or *Positive Penguins*. Taking care of our mental health is (finally) a priority for many of us, and children's is just as important (if not more). Well-being means also being physically active, and this is incredibly important for kids as it is for adults. Products like *GoNoodle* are designed with the intent of keeping kids away from a sedentary life, by also adding some fun to the mix. *Walkr* is a perfect example of gamification and fitness; it's a pedometer app where the measured steps fuel a rocket during a journey into space.

Video Games for Kids

In my introduction, I explained why I won't be talking about video games that are suitable for children but not specifically made for them. Normally games that are considered OK for children are actually made for a wider audience, including casual gaming adults. Mobile casual arcade games, such as *Angry Birds* or *Crossy Road*, are fun for a lot of people, regardless of their age. Video games specifically made for kids do exist, but they tend to include some kind of educational purpose; therefore, in my categorization, they would fit in the "educational" bucket, or in the sandbox or simulation categories described previously.

Digital Toys Are Not Video Games

> We all know how tough children are with toys. It turns out grownups are much worse.
>
> —Martin Cooper, American scientist

As I mentioned previously in this book, children's app developer Toca Boca seems to be the first to come up with the term "digital toy," or at least the one who better represents this idea. We already talked about how playing is a cross-species activity that seems to have a wider scope than just preparing young individuals to life as it was previously thought. Sure, that's one of the main goals when lion cubs fight each other, for example; they are learning how to silently sneak behind a prey and how to attack, along with learning a pride's dynamics. The same thing happens with little humans when they imitate adults (and therefore it is so important for us to provide them with positive role models to mimic) and why we have toys like little kitchens or vacuum machines or mechanic's tools modeled on the ones we, grown-ups, use every day.

These toys are tools that help kids in their "pretend" games. There is no winner or loser with these activities, just storytelling and experience. Kids playing "the kitchen" together learn about collaboration, about chores and responsibilities (tidying up, cleaning, buying groceries, etc.), there aren't any prizes, points, achievements, recognitions of any sort, the activity has a starting point, but not clear ending, it could go on for hours.

Such activities also involve a big deal of imagination. Toys can help this imagination, but ultimately it is the kid's creativity doing the work. My almost 3-year-old daughter often plays "the bookshop"; she loves books, and she has a lot of them, so sometimes she pretends to have a shop and I have to be the customer (even though, funny enough, I'm almost never allowed to choose what to buy by myself). She sets up shop by going behind an armchair that functions as the counter, when I (she) decide(s) the book I want she brings it close to a very specific part of the armchair and make a "beeeeeep" with her voice, as she if she scanned the book with a barcode reader, then proceeds to swipe an old Costco card between the armrest and the back of the seat to check me out, after putting the book in a bag along with my receipt.

For this activity, which could go on for tens of minutes, we just need

- An armchair (it could be anything else providing some sort of separation between her and the customer)
- An old plastic card
- Some books
- A bag of some sort
- A random receipt or a small piece of paper
- A lot of imagination and observation of adults' behaviors

We didn't buy any fancy toys, the cost is basically zero, the imagination effort is maximized.

Thanks to this role-play, she learns

- How shopping works
- What a payment card is
- How to handle various objects (swiping the card, putting the book in the bag, handing the receipt, reordering books on the shelf, etc.)
- Language skills (talking to a customer and pretending the customer is not dad or mom)
- Empathy (assisting the customer in finding a book, asking if they like the book, etc.)
- Tidying up

Much like these real-life playing activities, digital toys provide experiences (usually based on real situations and jobs) where there is goal other than the experience itself. There are no points, no levels, no matches, no winners and losers in apps such as the hits *Toca Hair Salon* (Figure 2-2) or *Toca Tailor* (Figure 2-3); creativity plays a big role in these products.

Digital toys can't, I repeat: *can't* be substitutes of real-life playtime, but can be a good variation and complement, and can stimulate creativity and imagination in a different way.

There are some advantages about the flexibility and convenience of digital experiences: we can play them in mobility (car, airplane, doctor's waiting room, you name it) and one device can store tens, if not hundreds, of different apps.

Figure 2-2. *Toca Hair Salon 4*

Figure 2-3. *Toca Tailor* Fairy Tales

Digital toys foster creativity and imagination. The lack of fixed objectives, rewards, and badges incites children into making their own pretend game, finding fulfillment in the process and in the end result of their creation, rather than in leaderboards, achievements, or points.

Where to Find Ideas

The difficulty lies not so much in developing new ideas as in escaping from old ones.

—John Maynard Keynes, English economist, The General Theory of Employment, Interest, and Money (1936)

You could develop a product because you have an idea and you think it might work on the market; maybe you never worked on digital products for kids before, but somehow got this inspiration and decided to give it a try, or maybe you already have experience in the industry and just came up with a new exciting concept. Awesome! But it could happen that the decision of getting into the kids' app market came because of a business opportunity, a partnership opportunity, or the request from a client, for example. But there is just the need of entering the market, without a clear concept in mind. This could be the case, for example, of a company already active in language learning apps for adults that decides to expand its audience targeting children. In this case, you have the expertise in digital products and in teaching languages, but no experience about doing so with children as users (and parents as customers).

In this case, you know the goal, but you don't know how to get there. You need ideas and inspirations to have a starting point.

Sometimes ideas come to you, unexpectedly, often in the craziest moments, and that's great. Other times they want to be looked for, they require exploration, scavenging, patience, and acute observation. You can find ideas anywhere, we all can agree on that, but let's consider the most likely scenarios where to look for ideas for this kind of products. Here's my compass, so that I won't leave you wondering around in the woods too much.

Within the Screen

Let me start with one thing I firmly believe: it doesn't matter if an idea is not new, as long as you can make it better.

If you think about it, this is the strategy behind a lot of Apple's products. Apple, in its history, has hardly been innovative, in a strict interpretation of the word. What Apple does is taking an idea that works but is poorly implemented and make it better from a UX standpoint, from an industrial design standpoint, and so on.

The GUI (graphical user interface, as opposed to text-based interfaces like MS-DOS) of the first Mac? Xerox already experimented with GUIs; in more recent times: the iPod? MP3 players were already a thing for quite some time. The iPhone? Smartphones and touchscreens existed already. And so on. So, you don't have to come up with new disruptive never-seen-before ideas for your product. But you don't have to copy either, that's obvious.

So, let's assume you come up with an idea, but then you find out something similar is already on the market.

The first thing you should think about is that finding other products similar to what you have in mind is an excellent form of validation and inspiration. That's right! If someone else went through the hassle of creating that product, investing time, money, resources in it, it's probably because your concept is targeting an actual need. It's something. It's actually more than something, it's what you need to know to really start investing in your own product.

Second, it saves time. Someone else researched and came up with a concept and then developed a product to solve that same problem/need you want to tackle. That doesn't mean you should skip the research phase, but checking existing products it's for sure an excellent starting point.

The point is not necessarily to be the first product on the market involving a particular concept. The point is being the one that brought that concept to life in the best way possible.

So, your idea already exists. What can you do?

- Look at all the products that could be your competitors. Not necessarily in the same exact market, think outside the box (banality alert!). If you're planning on making the new Spotify Kids, don't look just at Spotify and Apple Music. Look at video streaming, YouTube playlists, dancing apps for kids, and so on.

- List all the good things you find in those products as well as the pain points as a user, but also from a business standpoint.

Where to find users' pain points? Start from the product's reviews! Those are an invaluable source of information. The next questions then become: What could your product bring to the users as added value? How can you differentiate from the others? Maybe you can

- Add a feature lots of people are asking for and it's still missing on other products

- Target a different group (for age, language, etc.)

- Have a different UX (more/less playful, more/less minimal, ...)

- Use a different business model (onetime payment vs. subscription vs. free with ads vs. free with premium version vs. ...)

- And so on

I once worked with Colto, an app developer specialized in educational apps for kids. Their initial focus was making apps to teach languages, and I've worked on a few products with them; one of those is called "Eli Explorer." Eli is a little bunny that is able to fly by spinning his ears like a helicopter, thanks to this amazing ability he can explore a looping and apparently endless world, where he can discover new things, cute characters doing funny stuff and lots of Easter eggs. The main inspiration came from some "open world" exploration games for preschoolers, but we iterated on the concept by adding characters' voices to teach children over 100 words, like food, animals, vehicles, colors, counting to 10, and many, many more, in 10 different languages. We also added a lot of randomized situations, so that the adventure feels different every time. The game won the first prize as *Best App* at the *European Conference on Games-based Learning* in 2014, as well as a *Teachers' Choice Award* and a *Parents' Choice Award*. But most importantly, my daughter loves it! It's been amazing to see a product I designed when she wasn't even born yet, to delight her so much, and becoming trilingual (Italian, Japanese, and English) she can really benefit from the educational value of Eli.

Besides kids' products, there are millions of other apps out there. Sometimes a good idea could come by taking a look at products for adults and making a kid-friendly version of them. YouTube released its own version for children with YouTube Kids, Spotify did the exact same thing, and even Facebook released a kid's version of its popular messaging app. It's useless to say that not all products are suitable for a conversation to become children-friendly, but exploring apps that live outside kids' realm it's absolutely worth a shot.

There will always be space on the market for *good* products, because no design is flawless, there's always room for improvement. The thing to remember is: don't copy, don't steal, but improve, make it better, make it easier, make it more enjoyable.

Real Life

For most of human history, we lived our lives outside of the screen (and hopefully we'll continue to do so); therefore, there are lots of activities to explore that could be translated into a digital version. I'll take the risk of being very annoying and remind, once again, that this doesn't mean that digital products should take the place of the real-life counterpart of such activities.

Board games, card games, arts and crafts are some of the inspiration sources you can look at. A digital version of paper dolls, where kids can cut out and decorate clothing with their finger, can be a valid alternative to the real activity when the situation doesn't allow for scissors and paint, for example, on a flight. Plus a digital version could bring life to the kid's creations, characters could interact with each other and with the kids and react to touch, voice, or gyroscope sensors, for example, or brought into our world, thanks to AR (augmented reality).

This is just the seed for an idea inspired by a real-life old-school game; there are almost endless out there waiting to be reimagined and enhanced with the use of technology.

Books and TV shows can also be valid sources of ideas. They can serve just as inspiration or, if possible, become a real partnership with their creators providing them an additional touchpoint for their IP. As of mid-2020, the app developer Colto I mentioned in the previous section is pursuing this business opportunity, creating games based on books and TV series for publishers such as Highlights and networks such as Nickelodeon.

If you take a popular show for children, like *Peppa Pig*, for instance, and watch a few episodes, you'll find lots of opportunities for interactive experiences.

Lastly, don't underestimate the value of simple observation of kids playing. Understanding their way of thinking, their relationship with each other, the group dynamics, the way they interact with objects, you can learn a lot on what their needs are and how to best try to serve them.

Observation

Here we are. We were talking about observing kids playing and interacting with each other. If you have kids, you might already know a thing or two, but let's see how to set up a slightly more controlled environment. When you observe adults interacting with a product in a test environment, you usually need a well-defined protocol for the test session, you can alternate free exploration with tasks to test specific functionalities and features, you often use hi-tech instruments, like eye tracking to develop heat maps, EEG and ECG sensors to monitor brain and heart activities, plus screen recording and filming.

Designer Debra Levin Gelman[6] suggests, in her book *Design for Kids*, to have a much simpler setup when dealing with little users. A room with some age-appropriate toys would be enough, even better if you have the chance of observing them in an environment where they feel comfortable and safe, like their home or classroom, for example.

If you already know what the app will be about (e.g., music or art or maybe it's about vehicles or puzzles), another good suggestion is to fill the space with just toys and games pertinent with the subject of your product; this way, kids will focus only on the activities relevant to your research.

This is also valid for what concerns the age group your product will target (it would be a good idea to have this sort of decisions in advance, to set the boundaries of your exploration). I suggest to observe kids interacting with toys and other tools both in small groups (two or three) and alone. Both situations can be valuable to discover new use cases for your product. You might discover group activities you never considered, or ways to play alone you didn't think of. Playing together is an important part of children development, and parents surely appreciate toys and games that can bring kids together (without fighting).

---

[6]Gelman, Debra Levin. *Design for Kids: Digital Products for Playing and Learning*. N.p.: Rosenfeld Media, 2014. Print.

Industry Insight: Interview with Gregg Spiridellis

Gregg Spiridellis and his brother Evan founded StoryBots (www.storybots.com), a children's educational media franchise, that includes animated shorts, digital and printed books, digital products, and a popular TV show on Netflix. StoryBots won several awards, including three Emmys, two Annies, nominations at BAFTA and Webby Awards, and many more. Gregg's experience with digital products started a long time before StoryBots, when, in 1999, he founded with his brother JibJab Bros. Studios.

Rubens: What sparked your interest in creating an intellectual property for kids?

Gregg: My brother Evan and I started a company called JibJab in 1999 that developed content for the Internet. By 2011, we each had young children and lamented the fact that there was nothing on television that we were excited to sit and watch with our kids, the way our parents would watch *Sesame Street* with us. We decided to build something cool ourselves!

R: Was the business opportunity clear to you from the beginning? If so, did it go as planned or did you have to pivot at some point?

G: There were a few things that were clear. First, we had over a decade of experience producing short-form content and felt comfortable that those skills were transferable to the kids' space. Second, the iPad had just come out, and it was clear that it was going to be a game changer for how kids consume content. Third, YouTube was proving to be a massive distribution platform, and we were confident that we could use it to build a big, global audience for a new children's franchise.

R: When did the idea of making the digital products, such as the iOS app and StoryBots Classroom, come about? Did you see the educational value of the brand since the beginning, or is it something that you realized later on?

G: We started by distributing short-form videos on YouTube, specifically a 26-video series featuring the letters of the alphabet. Shortly after those videos were released, we started hearing from teachers about how helpful they were in the classroom. We realized that one teacher reached anywhere from 20 to 40 kids each school year and that there was no better brand ambassador to parents than their child's teacher. A few years into developing StoryBots, we decided to build a dedicated product for the classroom designed specifically for SMART boards.

R: StoryBots was a subscription-based service. It's a business model that gained popularity in the past few years. A lot of software companies are shifting from a onetime payment to a monthly or yearly subscription. Looking at reviews though, users often complain about the real value of these products. What's your advice to make users perceive the value for money in a subscription-based service? Frequent updates? Fresh content? Customer care?

G: All of the above! We had the most success with our subscription model when we were releasing new content every month and the users knew it. Specifically, when we had a digital "book of the month" club offering. That said, I will say that building a big business in the kids' space is very hard. There are a lot of competing free products vying for kids' attention (YouTube, Minecraft, Roblox, etc.), and acquiring new customers in the category is very expensive.

R: StoryBots, both the TV show and the related digital products, is an award-winning brand. Multiple Emmys, Annie Award, Webby, and BAFTA nominee… Did these many awards help in marketing the products to parents? Are awards a valid marketing tool in products for children?

G: I do not think general awards are particularly valuable from a customer acquisition and conversion standpoint. They help build the glow around the brand, but I do not think a parent will be more likely to pay for a product based on the winning of a general entertainment award, like an Emmy. That said, there were some awards, like the Teachers' Choice Award, that I think helped validate the educational value of our products and that always help convert parents who are free users into paying customers.

R: One thing I love about *Ask the StoryBots* TV show is how it's made to entertain kids and parents at the same time, something that most TV shows for young children can't achieve—I can speak from experience. One thing that is often missing in digital products for kids is this shared moment of fun. Is there a secret sauce to capture the interest of such diverse users at the same time?

G: Creating shared experiences is so important! As a parent, there is nothing that you value more than time spent having fun with your kids. It was the main goal we set out to achieve when we started producing *Ask the StoryBots*, and it was the key driver of some of our most successful apps.

R: Last question. We talked about children and parents, but another big piece of the puzzle is educators. How can we get educators on our side when developing a digital product for children? What are they looking for in such products?

G: The best way to get educators on your side as you are developing new products is to invite them into the development process. While that is hard at scale, when you talk to educators, you will get tips on how your product can make their lives easier. If you can make a teacher's life easier and bring some fun and excitement to the learning experience in their classroom, you have the ingredients for a successful product!

Chapter Recap

- Digital products for children can range from entertainment to educational, and every shade in between. Your product does not need to be educational at all costs, just fun is also good.

- We can identify different big buckets of digital products for children: educational, simulation, sandbox, content library. Most of children's products currently on the app stores fall into one of these categories.

- Digital toys are interactive experiences, similar, in a way, to video games, with the big difference that they do not offer a specific challenge, they don't have a leaderboard or points. They are toys where the storytelling and the goal are defined by the kid playing freely.

- You can find new ideas in different ways: looking at the market. What's out there? How can I make this better? You can look for inspiration in the real world, or in books and TV shows for children; how can I make this story interactive? Lastly, observation of kids playing is a great way to identify new design opportunities.

Know Your Target Audience

Know Your Users, Their Needs, and Expectations. Understand Challenges. Find Opportunities.

> *I love kids, but they are a tough audience.*
>
> —Robin Williams, comedian

We briefly introduced, at the beginning of this book, how the products we're going to design will have to deal with three different kinds of audiences: children, parents, and educators. In this chapter, I want to go further into this analysis and highlight the differences and the reasons why it's imperative to acknowledge all the three since the concept phase of the project development.

A Threefold Audience

Normally any product, digital or not, involves a reference target. The target is the answer to the question: who is this product for? It can be a group defined by age, gender, language, territory, habits, by tastes, and many more segmentations, also in combination to one another. For example, we aim a product at people 25 to 34 years old, living in big cities, who love doing sports and listening to music. That's a broad target. By adding more filters, we can shrink it down depending on our business strategy.

Usually, a product has a broader or narrower target audience, but it's rarely more than one (with the exception of products/services heavily focusing on content, where the variety in genres can please many kinds of users, e.g., streaming services). Kids' products, especially the digital ones (and especially the ones requiring a subscription), have different kinds of users to please, and these users, more often than not, have very different goals in mind. As you already understood, we are talking of children and parents, and to these two categories I like to add educators, I'll explain why soon.

In this Venn diagram (Figure 3-1), one circle represents the interests and goals for the kids and another the parents'; we can see how small the intersection of these two is, while the goals for parents and educators overlap much more (but not completely; teachers might care more about collaboration features, while parents monitoring screen time, just to make an example).

Kids' goals

Parents' goals

Educators' goals

Shared goals

Figure 3-1. The needs' overlapping of kids, parents, and educators

This is a key distinction between products for adults and products for kids that makes the latter more challenging to market. Normally kids have more interest in the fun part of the product, while parents care most about the educational value. Therefore, in Chapter 2, I remarked how crucial it is to find a good balance between the two. But you don't have to think about educational value in a strictly academic meaning. A game à la *Angry Birds* can demonstrate about physics of bodies, gravity, and so on; it's an entertaining game to play, for both kids and adults, but while grown-ups should already have those concepts clear in their mind, for children this could serve as a lesson and not just an occasion of mere entertainment.

The Kids

This is our primary focus, because as designers we share with the parents and the educators the ambition to create something valuable for children.

By "valuable" I mean

- **Not harmful (physically and psychologically)**

    The product must comply with laws concerning children's safety and with best practices as advised by pediatricians and other experts.

- **Focused on their development**

    The product should contribute positively to the kid's development, both cognitive and physical. Regardless of the main scope of the app being education or not, it should always provide an experience where the well-being of the young user is at the center.

- **Fun to use**

    Nobody enjoys boring stuff, and this can't be truer for children. Especially when your product's intents are educational, you want children to have fun when using it, to keep their interest and curiosity alive. It's like making a medication sweet in order for them to take it. There's no purpose in a bitter syrup that no kid wants to swallow.

- **Easy to use**

    This is strictly connected with the previous point. Even the most fun idea becomes boring if the app or website is burdensome to use. You don't want to frustrate your users, losing their attention is extremely easy. Adults can endure the burden of a terrible UX because maybe they have no other choice (e.g., with software related to their work), but kids want to use these apps for entertainment and edutainment. Don't spoil the fun with a bad UX.

- **Accessible and inclusive**

  Diversity has finally entered design talks. With our communities being more and more multicultural, even if you work on a product for a specific region, you want to consider inclusivity as an added value. Besides ethnicity or religion or home country, diversity can also mean being affected by physical and/or cognitive impairment; inclusivity means taking this into account and working on accessible products.

In the introduction of this book, I briefly pointed out how kids, compared to adult users, present a segmentation based on the age. This can be a problem, because, as you can imagine even without being a parent, a 3-year-old has very different cognitive abilities and motor skills than a 10-year-old. The first one most probably cannot read, and the kid has a much more limited knowledge of the world, a limited ability for abstraction, different logic and mathematical skills, but also a different dexterity when it comes to performing gestures, a *drag-and-drop* action that can be very easy for the older kid could be tricky for the younger one, and so on.

This book cannot address all the details of physical and mental development of children; that's a task for medicine manuals and pediatricians. What I can do though is giving some input and more practical tips, and we'll see these later when going deeper into UX and UI matters.

Depending on the age of the kid, the parents also have a different weight when it comes to decision-making. For younger kids, which apps to spend money on and download is a parent's choice, while older kids might have a bigger influence on such decisions. So, it's still true that when designing you should always have both children and parents in mind, but the balance shifts a little as kids grow up.

So, as you can see, our first audience is already fragmented on so many levels. It's important to understand and decide which age group we design our product for. I used the singular "age group" and not "groups" on purpose; considering the sizable differences in development occurring in even just 1 year, it's a tough job to make a product suitable for more than a 2–3-year span. Common age groups are 3–5, 6–8, 9–11 or sometimes 2–4, 5–7, and so on; on both Google Play Store and App Store, they call the youngest group "5 and under."

The Parents

As we said earlier, a parent's role as decision-makers, when it comes to buying digital products, usually decreases as the kids grow up. But no kid is economically self-sufficient; for this reason, you still have to hold parents' needs (and opinions) in high regard.

The needs of the parents are different from their kids'. They care more about the educational value an app might have, how safe it is, how addictive it is. Is it something they could leave their kids alone with for a few minutes, or does it require constant supervision (hint: if it requires constant supervision, something doesn't seem right)?

Plus other considerations of financial nature, such as the price of the product or, depending on the business model, cost of the subscription, and consequent value for money relation.

Whether a product is appropriate for children or not can be, in part, very subjective. There are boundaries determined by the law, but for minor things like poo or fart jokes, things are interpreted very differently, because of culture or simply personal taste and beliefs.

Parents hold two powerful bargaining weapons that can make or break our products: the wallet and the reviews on the app stores (or any other platform featuring user reviews).

To spend money on a product, parents need to be convinced about its quality, value, and safety. In this book, we'll cover several topics to achieve all of those, from a design and from a marketing standpoint.

The reviews can be a curse or a blessing for any kind of product, but parents, being naturally (and rightfully) protective over their children, have no mercy when a product is below their expectations, so pay attention to the reviews.

Parents usually trust educators, so it's important to have them on board.

The Educators

Teachers are our best allies. They share two things with the parents: they care about children and they are adults. But teachers have different needs from parents and also a different spending power. Especially in public schools, teachers might not have enough funding to buy your product in bulk, so consider giving free access to them. Why? Because teachers can become your ambassadors to promote the product among parents (and other teachers, who will do the same with more parents in a potentially viral mechanic).

Teachers need to have genuine reasons to endorse your product, so quality matters even more with them. While parents might not have the right tools to evaluate the quality of an app for kids, teachers know when a product is worth being used in classrooms and promoted for use at home. Some teachers also have blogs and social media channels where they give advice on digital products to peers. So it's important to involve teachers in the development process from the very beginning, having an education expert consultant in the team is of paramount importance, as it is having a pool of beta-testing teachers to collect feedback from.

When you design an educational product for children and you expect it to be used by teachers in classrooms, know that teachers will require a back-end dashboard where they can analyze the usage of the product, the results, keep track of activities, possibly customize such activities, and more. This part will be a different task from the kid-facing side of the product as it will be a part of your children's product designed for grown-up needs in a professional setting, and this is just another example on how complicated a product "for kids" can be.

Cross-Platform Example: ClassDojo

ClassDojo is a good example of a cross-platform product that is made for kids, parents, and teachers (Figure 3-2). Its purpose is to be a classroom communication platform, to connect teachers, students, and families. To make a few examples, teachers can share with parents moments of classroom life, kids can see their assignments and get grades and comments from the teacher, and parents can ask questions and get in touch with teachers as well as monitoring the progress of their kid.

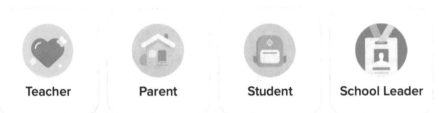

Figure 3-2. Onboarding for different users in ClassDojo

Each kind of user gets a different experience. For example, here are the different dashboards for student and teacher (Figures 3-3 and 3-4).

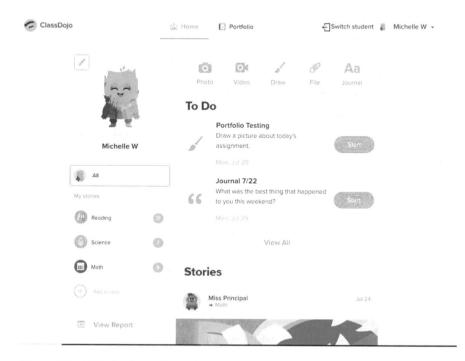

Figure 3-3. Web dashboard for students

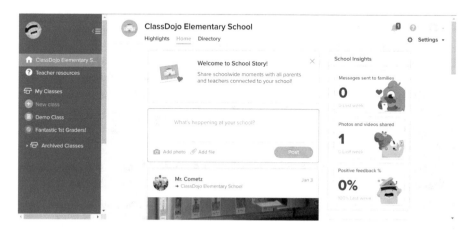

Figure 3-4. Web dashboard for teachers

In this case, the product is probably slightly more aimed to help teachers in organizing activities and engaging children and parents. So, it could be that your product for children doesn't have kids as its primary target.

When I was working for StoryBots, we had two main products. One was the iOS app that was aimed to children first, a container of videos, books, and activities for preschoolers. The second one was StoryBots Classroom that was a digital platform for tablet, PC, and SMART board to use at school and at home, to teach math in an engaging and fun way. So while with the iOS app the focus was mainly on kids, with StoryBots Classroom we shifted more toward the educator's needs (teaching, engaging, organizing, etc.).

Explore the possibility of serving the needs of parents and educators first and the kids' as a consequence of that. Sometimes this change of perspective helps come up with new ideas and opportunities.

Industry Insight: Interview with Scott Emmons

Scott Emmons is an author and songwriter. He wrote several works for children, including songs and bits for the Netflix TV show *Ask the StoryBots*, early readers' books, and educational songs for animated shorts.

Rubens: I had the good fortune of seeing you working on StoryBots' TV show, books, and songs, and I was amazed by the creativity and the fun you put in everything you wrote. The first question might sound banal, but I'm curious to know where you find your inspiration.

Scott: Well, thanks for the kind words! I'd say I get the most inspiration from reading, watching, or listening to creative work that I enjoy and admire. I do a lot of rhymed verse writing for kids, so I'm inspired by real craft masters like Dr. Seuss, Shel Silverstein, Jack Prelutsky, and so on. Ideas can also come from just paying attention to things that happen and seeing if you can make something fun out of it. When I sit down to write, I often start with some brainstorming, just free associating until I find an idea I want to play with. I'm also a big believer in tapping into the unconscious by taking a walk or doing some other activity and letting ideas bubble up.

R: You write content for grown-ups as well. What are the main differences between how we speak to kids compared to adults?

S: For me it's not as big a difference as you might think. One of my main rules in writing for kids is not to talk down to them. Even if they don't necessarily understand every word, engaging content will hold their interest. That was one thing that was so rewarding about working on StoryBots. They serve up some really challenging concepts in ways that make them fun for kids. At the same time, I'd say my writing for kids tends to be a bit more whimsical. Kids are more willing to embrace nonsensical ideas, so if I want to write about a dinosaur with bunny ears, I can just go for it.

R: All StoryBots' content is super fun, and educational as well. What's the secret to combine these two aspects? They usually are not easy to put together. What's the secret?

S: Again, I think the secret is to include a touch of whimsy. I didn't write the Outer Space song series, but that's a great example. What better way to teach kids about the celestial spheres than with a rapping Sun, Moon, and planets? I did write a series of bluegrass songs about numbers 1 through 10. For that, I could often just riff on fun things that would be familiar to kids. Like for number 3, I could bring in the three little pigs, the three little kittens, Old King Cole's fiddlers three, and so on. It's fun combining education and entertainment for kids, because just about anything can be a character.

R: StoryBots has been able to make content for kids enjoyable by adults too. Do you think your experience with writing for grown-ups helped you in writing for children while winking at adults?

S: It goes back to the point about not talking down to kids. *Ask the StoryBots* is full of little jokes and asides that are over kids' heads, but the attitude is still fun for kids, and the adults can enjoy the added bit of humor. Sometimes you can do something that operates on more than one level. I wrote the amphibian song for the "Animals" episode. The character singing it was an obvious nod to Michigan J. Frog from the old Chuck Jones cartoons. Most kids probably didn't get the reference, but they could still enjoy it just as a fun song.

R: You wrote a series of books for early readers. What is important to keep in mind when writing this kind of work? Things like shorter sentences, or avoiding passive voice? I'm making assumptions here. You're the expert!

S: Those "Step into Reading" books look very simple, but they were some of the most challenging things I wrote for StoryBots. Yes, there are strict guidelines for word and sentence length. When I write rhymed verse, I'm very meticulous about meter—the rhythm, accentuation, number of syllables, and so on. It just about killed me that I wasn't allowed to use contractions. That added a whole new layer of difficulty! But the books have been really successful, so I'm not complaining.

R: Many of the topics that the StoryBots present, in the show, in the books, and in the short videos, are scientific facts that seem very hard to explain to young children. But they do it in a very successful way. What's the secret here?

S: I have to give credit to the Spiridellis brothers (creators of the StoryBots) for not holding back on those complex subjects. They don't dumb things down! It's often a matter of breaking down the basic concepts into understandable bits of information that kids can digest. I did a lot of songwriting, and I think songs have the added advantage that kids will listen to them again and again, so the information gets reinforced. The most challenging assignment I had was to explain gravity in a one-minute song. A lot of the information probably zipped right by the kids the first time they heard it, but with repetition it would stick. I hope so, anyway.

R: You have a past as a Greek and Latin teacher. Even though you weren't teaching to children, do you think this experience helped you in creating educational content?

S: To be honest, I never really thought about it. Maybe it helped in some intangible ways. I talked earlier about breaking down information into digestible chunks, and that's a big part of teaching. I think my academic background helps in some ways because it gives me a deep understanding of words and language. On the other hand, I sometimes have to shake myself out of my academic mindset to get to the really fun stuff.

Chapter Recap

- Designing for kids means designing for parents and teachers as well.

- Kids put fun in front of other needs.

- Children's audience is split in age groups. Each group has different needs and presents different opportunities and challenges.

- Parents can be a very demanding audience. They want the best for their kids and pretend the best value for money.

- Educators can be our allies, if they trust our product and find real value in it. They can become ambassadors for our product and promote its use to parents and other educators.

- A product trusted by educators looks more reliable to parents.

Concept

Set the Groundwork for Your Product.

This chapter is dedicated to understanding the decisions we should take before actually going to the whiteboard and start brainstorming ideas. In some cases, some people might decide to take such decisions along the way, but, in my opinion, it's better to have at least some fixed points to define a perimeter. This speeds up the process.

Decisions like: Which devices work best for my target audience? Should this product be a native app or a web app? What kind of experience is this going to be?

Having at least some of these questions answered gives us some bricks to build the foundation of our concept. It doesn't mean you should feel stuck and never change idea on any of them. You can, and in most cases you will, but at least you won't suffer from blank page anxiety and you'll start with a more streamlined process.

How can we take these decisions beforehand? Research and strategy. In the previous chapters, we looked at the industry (competitors, business opportunities, etc.) and the audience(s). From this we should define what our opportunities are and how we can make the best out of them. Let's say that we got a hint from talking with elementary school teachers, about a pain point they have with 1st and 2nd graders. Which are the devices on which 5–7-year-olds should use our product? What kind of problem is this? Is it about a specific subject? Is it about engaging the whole classroom? Is it about reading? Or math?

© Rubens Cantuni 2020
R. Cantuni, *Designing Digital Products for Kids*,
https://doi.org/10.1007/978-1-4842-6287-0_4

If we identify a problem to solve, a need that is unfulfilled or not fulfilled properly by other products, we can answer many of the preceding questions and have the groundwork to brainstorm ideas.

Which Technologies Should We Use?

So, we're finally here! We studied the market; we learned about our audience(s) we made our research on competitors and now the team is ready to start exploring possibilities and coming up with ideas.

Though there's one important decision we should make before starting with ideas for the product, and considering how this decision will influence the outcome of the concepts we'll define, it's better to have this clear in advance. The question is: What kind of product do we want to make? For which devices?

The choice between a responsive website, a native app, or even a physical product with a companion app will influence the limitations and the opportunities, both on a technical side and a marketing and monetization side.

What are the options? Well, as they say: sky's the limit. The most common and obvious choices are mobile app and responsive website/web app. Let's take a look at the key differences between the two.

Mobile Apps

A native app could be either for Android or iOS (at the moment I'm writing, these are the only two options available for mass market). You can, of course, develop and release a version for each operating system; just keep in mind that, even though Android devices are currently the majority, statistically iOS performs better in terms of revenue per user.[1]

Unless your team is big enough, you'll most probably need to develop one of the two first, so where to start? Which is better?[2]

---

[1]Perez, Sarah. "App Revenue Tops $39 Billion in First Half of 2019, up 15% from First Half of Last Year." *TechCrunch*. TechCrunch, 03 July 2019. Web. 31 May 2020. <https://tech-crunch.com/2019/07/03/app-revenue-tops-39-billion-in-first-half-of-2019-up-15-from-first-half-of-last-year/>.

[2]Aggarwal, Shanal. "Android vs IOS - Which Mobile Platform Is Better in 2020?" *TechAhead*. N.p., 18 Jan. 2019. Web. 31 May 2020. <https://www.techaheadcorp.com/blog/android-vs-ios/>.

There are several considerations to take into account; let's see some data to have a clearer picture.

- Android is the leader by far in terms of market share, with about 75%.

- Revenues on iOS though are almost double of Android's (14 billion dollars vs. 7).

Table 4-1 shows a comparison between iOS and Android apps, both in terms of revenues and downloads.

Table 4-1. iOS vs. Android comparison in revenues and downloads in 2019[3]

	iOS	Android
Revenues	$14.2B	$7.7B
Downloads	$8.0B	$21.6B

As you can see from Table 4-1, with about one-third of Android's downloads, iOS got almost double the revenues! That's pretty impressive. It's clear that iOS users are more willing to pay for apps and subscriptions than Android's. This is also due to Android's open source nature that made the birth of alternative stores and bootlegged versions possible. Apple's closed and controlled ecosystem makes digital products for iOS less susceptible (though not immune) to piracy and illegal exploitations.

Now, let's consider the technical aspects of developing for one platform or the other.

- **Development complexity**: Android's ecosystem is way more fragmented than iOS. Despite having added a lot more screen sizes in recent years, the number of screen ratios and resolutions you have to deal with when developing for Android is considerably higher than iOS'. Lots of different brands with an endless catalogue of models compared to a handful of devices (at the moment I'm writing there have five different iPads and five different iPhone models).

---

[3]Data from: Chapple, Craig. "Global App Revenue Grew 23% Year-Over-Year Last Quarter to $21.9 Billion." *Sensor Tower Blog*. N.p., 23 Oct. 2019. Web. 31 May 2020. <https://sensortower.com/blog/app-revenue-and-downloads-q3-2019>.

- **Development time**: Statistically[4] developing apps for Android is 30 to 40% slower than doing so on iOS.

- **Development cost**: The estimate here is complicated, because while a longer development time could mean a higher cost, the reality is that Xcode, the development tool used to make iOS apps, runs only on Apple devices; therefore, you would need an expensive MacBook to make apps for iOS, compared to a much cheaper Windows PC to make apps for Android (be aware that this doesn't go both ways: you absolutely can make Android's apps on a Mac).

- **Programming languages**: The official language to build Android apps is Java; in 2017, Kotlin has been added as secondary official language. C++, C# can be used with Android Native Development Kit (NDK), and by using tools to convert into Android Packages, even Python could be an option. Furthermore, Android apps can be created using HTML, CSS, and JavaScript using the Adobe PhoneGap framework that is powered by Apache Cordova, but these will be basically web apps shown through *WebView* and packaged like native apps. iOS official languages are Objective-C and Swift. All of this means that you will need people skilled in at least one language for each platform or different developers to make apps for both systems.

- **Publishing process**: Apple has a quite lengthy and restrictive process to upload and approve apps before going to the store, while Android's is faster and easier. Approval time too usually takes longer on iOS, and this reflects also on updates, which are almost immediate on Android while for iOS is required again to go through the approval process. Cost-wise, the fee to become developer on Google Play Store is a onetime payment of $25, while Apple asks $99/year for the privilege of being on the App Store.

[4]Aggarwal, Shanal. "Android vs IOS - Which Mobile Platform Is Better in 2020?" *TechAhead*. N.p., 18 Jan. 2019. Web. 31 May 2020. <https://www.techaheadcorp.com/blog/android-vs-ios/>.

Taking all of this into consideration, the decision ultimately depends on your priorities. If your priority is the profitability of your product, you probably want to go with an iOS version first, easier development, maximized profits. If you care more about reaching the largest audience possible, because it's a nonprofit project, for example, then Android is the way to go.

Design-wise, generally speaking (and take this with a grain of salt), products on iOS are superior in terms of user experience and visual design. I don't think we can really pin down a single reason, but my guess is that this is due to a combination of factors. Being a closed ecosystem where the rules to go live on the App Store are much more strict than the ones on Google Play Store created a much more curated and high-quality catalogue of apps; therefore, competition is set to higher standards and you want to be up to par, or possibly better, with them.

Android's unrestrictive approach created the perfect playground for hordes of indie developers and hobbyist who blessed us with so many great products that let us do things impossible to achieve on an iOS device, but also bloated the ecosystem with low-quality, possibly (or intentionally) harmful, unpretentious products that set the bar much lower than iOS'.

Ultimately people want to pay for quality or not pay at all; it's easier to ask for money when your product is done right and looks right.

The *aesthetic-usability effect*[5] is the "**tendency to perceive attractive products as more usable**. People tend to believe that things that *look* better will *work* better — even if they aren't actually more effective or efficient."

Combining this with the ease of downloading Android's apps illegally, you can understand why making products for iOS is far more profitable.

Web Apps

Web apps are displayed in a browser and requires an active connection in order to work. Modern web apps are *responsive*, meaning that they adapt their layout to whichever container they get displayed in. There usually are *breakpoints* that dictate the behavior of the components of the app according to specific size intervals, so this defines how the layout differs when you display the web app on a smartphone compared to when you do the same on a tablet or the bigger screen of a computer. This is very different from native

[5]Moran, Kate. "The Aesthetic-Usability Effect." *Nielsen Norman Group*. N.p., 29 Jan. 2017. Web. 31 May 2020. <https://www.nngroup.com/articles/aesthetic-usability-effect/>.

mobile apps, where you normally still have a certain degree of responsiveness to allow your app to adapt from an iPhone 8 to an iPhone 12, for example, but you need a specific build for tablets.

An even bigger difference is that web apps are *platform agnostic*. This is a big deal because being platform agnostic means that you don't need a different app for iOS and Android and even for computers! As long as the browser is supported, your app can be displayed (yes, there might be some limitations, for example, from computer to smartphone, but you won't need a completely different app, built with a different technology on each platform).

In Figure 4-1 you can see how Medium is different in its Android native app (left) compared to the web version (right) and the iOS app is again different from the Android's.

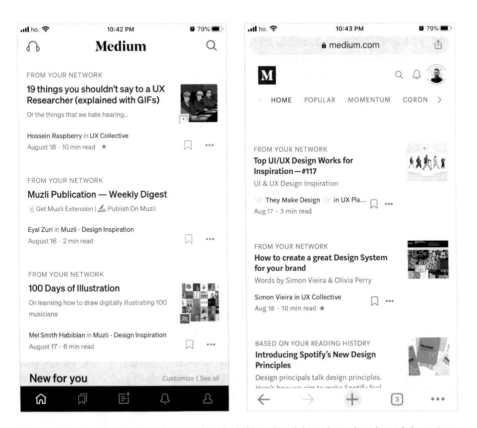

Figure 4-1. Medium in its native app (on the left) and mobile website (on the right) versions

Native apps are usually faster and give a more *snappy* feeling in their user experience, animation runs smoother, and the overall experience is more immersive. But let's make a list of pros and cons of both worlds (Table 4-2).

Table 4-2. Mobile apps vs. web apps

	Pros	Cons
Mobile apps	Faster	More costly to build compared to web apps
	Can work offline	
	More functionalities as they have better access to the device's hardware	They usually require a specific build for each platform
	Safer. They require approval to appear in stores	Expensive to maintain and update
		Updates need to be installed on the device
	Easier to find, thanks to the app stores	
	Not running in a browser (much safer for kids)	Sometimes difficult and inconsistent app stores approval process
	Can send push notifications	
Web apps	They function in browsers, no need for installs	Do not work offline
	Platform agnostic, they don't need a specific build for each device	Slower than native apps and often lacking some features
	Self-updating without any local install or intervention from the user	Less discoverable as there is no store
	Quicker and easier to build than mobile apps	The user experience is usually inferior compared to a native app
	They don't need approval as there are no official stores, they live on the Web	

Progressive Web Apps

The term Progressive Web Apps was introduced in 2015 by designer Frances Berriman and Google Chrome engineer Alex Russell. They defined a series of requirements for an app to be considered a PWA (*Progressive Web Application*). There is still a bit of controversy around what a PWA really is, it seems to be sitting in a gray area where the perimeter to actually declare when a web app is a PWA is still blurred, but anyway, the whole idea is that PWA should incorporate the best of both worlds.

Let's see the requirements according to Google Web.dev[6] and other sources:

- **Progressive**: Works for every user, regardless of browser choice.

- **Responsive**: Fits any form factor—desktop, mobile, tablet, or forms yet to emerge.

- **Faster after initial loading**: After the initial loading has finished, the same content and page elements do not have to be re-downloaded each time.

- **Connectivity independent**: Allow offline uses or on low-quality networks.

- **App-like**: Feels like an app to the user with app-style interactions and navigation.

- **Fresh**: Always up-to-date due to the service worker update process.

- **Safe**: Served via HTTPS to prevent snooping and ensure content hasn't been tampered with.

- **Discoverable**: Identifiable as an "application" by manifest.json and service worker registration and discoverable by search engines.

- **Re-engageable**: Ability to use push notifications to maintain engagement with the user.

- **Installable**: Provides home screen icons without the use of an App Store.

- **Linkable**: Can easily be shared via a URL and does not require complex installation.

Put like this it would seem like PWA are the obvious choice, and that's why a few years ago many predicted that PWA would have killed the native app star. But that didn't happen, at least for now and for a foreseeable future.

What is the difference with native apps then? There are several things to consider, because while it's true that PWA combine many pros of native and web apps, it's also true that they present some of the drawbacks. For example, the lack of security given by an approval process, anyone can create a PWA and release it online, because, other disadvantage, there is no store.

Without diving down into technicalities, because this is not the book for that job, PWA at the moment seems better on paper than they are in reality.

---

[6]Google. "Progressive Web Apps." *Web.dev*. N.p., n.d. Web. 31 May 2020. <https://web.dev/progressive-web-apps/>.

So, What Should I Choose?

Ultimately native apps provide more security, which is very important in kids' products, a better user experience (smoother transitions, faster response …), and a better ROI (return on investment) in terms of branding and credibility, something you shouldn't underestimate when you deal with parents.

Among native apps, based on the data presented before, iOS is the platform where you can get higher profits, simply put, while Android more downloads, if that's what matters to you.

Choosing the Right Device According to Kids' Age

We talked about native apps, web apps, and PWA, but on which kind of device we should run these apps is an even more important question. During the early stages of our concept development, we need to decide: Which device(s) should our app run on? Smartphones? Tablets? Computers? All of them? This also influences greatly the way your users will interact with the app.

Each device has peculiarities that go from screen real estate to the way users interact with it. A smartphone or a tablet is mainly a touch-based device, while a computer's main way of interaction is through a combination of keyboard and mouse or trackpad. But smartphones and tablets also offer more sensors, like the gyroscope and the accelerometer, for example, and these can be exploited to develop alternative ways of interacting with the device.

Human-computer interaction (HCI) is heavily influenced by the degree of ability of the user. For a person affected by Parkinson's disease, for example, pointing tiny targets with a mouse can be a very frustrating experience. Our body develops in stages where we progressively refine our dexterity and finesse in handling objects; newborn babies have uncontrolled movements as their muscles and nervous system are still "under construction." As the baby grows, she gets more and more accustomed to her body and gains a higher degree of control over it, but this process goes on for a few years. Toddlers are still quite clumsy with their hands, not to mention how tiny those chubby little hands are. Therefore, the way a 3-year-old interacts with a device is very different from what a 6-year-old can do.

Let's take a look at the stages of physical development. Starting with motor skills, we can define three levels[7]:

- **Gross motor skills**: Movements that involve large muscle groups, like those in arms or legs, for example, actions like jumping, skipping, and so on.

- **Fine motor skills**: Precise movements that involve the small muscles in the wrists and fingers. Writing and drawing with pen on paper, grabbing a small object between thumb and index finger....

- **Motor coordination**: The general ability to combine the two and coordinate different parts of the body in order to accomplish a particular task.

These three levels highly vary from one age group to the other; consequently, the preferred gestures and devices change accordingly, as you can see in Table 4-3, based on an article by the Nielsen Norman Group (NNG).

Table 4-3. Physical development in children based on age and suggested devices (Data source: NNG)

		3–5-year-olds	6–8-year-olds	9–12-year-olds
Physical ability	Gross motor skills	Limited	Partially developed	Well developed
	Fine motor skills	Very limited	Limited	Well developed
	Motor coordination	Very limited	Limited	Partially developed
Device preference		Touchscreens	Touchscreens and PC	Touchscreens and PC

From the table, we can see how simple gestures such as tap, swipe, and drag are easy even for younger children. These gestures come natural for most users, regardless of the age as they mostly involve gross motor skills that, in the majority of children, develop earlier than fine motor skills. Of course, tapping interaction needs bigger targets for younger children, for the reasons explained earlier.

Interaction with a mouse is more complicated to handle, trackpads seem to be of help for some kids, but in general we should consider involving the use of a mouse on products for kids that are 9 and older.

---

[7]Liu, Feifei. "Design for Kids Based on Their Stage of Physical Development." *Nielsen Norman Group*. N.p., 8 July 2018. Web. 31 May 2020. <https://www.nngroup.com/articles/children-ux-physical-development/>.

It's worth noticing though that in tests conducted by NGG, even older kids who were able to use a mouse abandoned it in favor of a touchscreen if prompted with such option.

Clearly, as for all of us, the level of dexterity is individual, so there might be kids who can perfectly operate a combination of keyboard and mouse at 7, but we can get to the conclusion that while touchscreens are a valid option for all ages, more traditional computer interactions involving keyboard and mouse should be considered for kids that are above 9 and older.

Touchscreens though can be of many different sizes. We can still find smartphones with a 4.7 inches screen (although rare by now) and we go up to an iPad Pro with its 12.9 inches. Considering the information in Table 4-3 regarding gross and fine motor skills, we can understand how bigger targets are better; we could even simplify by saying that the younger the kid the bigger the target.

It's easy to see how tablets should be preferred to smartphones, when it comes to touchscreens.

Now, if I had to draw a conclusion from all we discussed in this section, considering business opportunities, security of the platform, and kids' physical development, I'd say that a native app for iPad is probably the first choice to go with. But there might be other considerations specific to your idea or business that won't make this the optimal choice. Hopefully, here, I've been able to provide to you enough data and information to make an informed decision.

Consider the SMART Board

Classrooms have changed a lot from what we remember (at least my generation and the previous ones). Several new devices are now in educators' toolbox, such as the ones we talked about in the previous section. But there's another device that is peculiar to the classroom and we don't have at home: the SMART board. What is it? How does it work? How can we benefit from it with our product?

A SMART board (which is actually the proprietary name of a product by SMART Tech), or an interactive whiteboard in general, is a big interactive display with the form factor of classic whiteboard. It can be in the form of a very big touchscreen or a combination of a whiteboard connected to a computer and a projector. Users can perform the usual gestures like tap, drag and drop, swipe, etc. and write with (specifically made) markers just like they would on a regular whiteboard, with the addition that what they draw can be interacted with.

As many pieces of our technology, interactive boards were first developed at PARC (once known as Xerox PARC, the place where the first GUI and mouse were invented) in the early 1990s, but they became popular as a classroom aid in more recent years.

Nowadays, many classrooms, from kindergarten and above, feature a SMART board, since 2007/2008 the popularity of this device within schools increased dramatically.

Their popularity is due to the fact that these interactive boards can do the job of whiteboards, touchscreen devices, projectors, and TV screens all in one. Teachers can use them for many activities, and differently from having touch devices for every single student or small groups, activities on the whiteboard can engage the entire classroom. Some manufactures also provide remotes through which students can collectively participate in polls and quizzes appearing on the whiteboard.

The first and most obvious benefit of an interactive whiteboard is that its use makes lessons more engaging. Combining animated visuals with sounds and interactivity is a great aid in keeping users, especially the young ones, engaged and interested. Teachers can also experiment with different teaching styles to meet the different needs of the students; in fact, we don't all learn in the same way, some of us prefer visual materials, others benefit more from touching and interacting, and so on, the versatility of the SMART board can offer many options to teachers.

SMART boards are essentially computers. As such, they give access to an endless catalogue of learning materials (because, you know, the Internet is quite big). Teachers can start an activity and then contextually search for supplementary material related to the subject.

And not just that, just as computers, they can be connected to a wide array of external peripherals, think about a microscope, for example, that can be used to enhance the learning experience.

Studies[8] demonstrated encouraging results in the use of SMART boards in classroom, for basically all K-12 children, resulting in more engagement, more

---

[8]Preston Chris, Mowbray Lee, "Use of SMART Boards for teaching, learning and assessment in kindergarten science." *Teaching Science* 54.2 (2008): 50-53. Print.

Hamdan, Khaled, Nabeel Al-Qirim, and Mohammad Asmar. "The Effect of SMART Board on Students Behavior and Motivation." 2012 International Conference on Innovations in Information Technology (IIT) (2012): n. pag. Print.

Jelyani, Saghar Javidi, Abusaied Janfaza, and Afshin Soori. "Integration of SMART Boards in EFL Classrooms." International Journal of Education and Literacy Studies 2.2 (2014): 20-23. Print.

Kırbaş, Abdulkadir. "Student Views on Using SMART Boards in Turkish Education." Universal Journal of Educational Research 6.5 (2018): 1040-049. Print.

motivation, and deeper understanding on a wide range of subjects, from language learning to sciences. This increase in motivation goes beyond the students, involving also the teachers enabling a virtuous cycle, where motivated teachers engage students and engaged students enhance teachers' motivation.

Designing for Interactive Boards

The ratio of most interactive boards is similar to a regular desktop computer (16:10), which is comparable to a tablet held in landscape orientation, but of course the size is much bigger. For this reason, you have to consider carefully the positioning of the elements on screen.

While an adult (with no disabilities) can usually reach with relative ease the top of the board, a child could not be able to. Some interactive boards can be moved up and down as needed, but sometimes are fixed in place, then it's important to consider the reachability of our interactive components. This changes according to the age group you're designing for, but it's important to consider also children with disabilities, who might not be able to reach as high as their classmates.

On interactive whiteboards you can run any software that can run on your computer, because they are basically another screen (with touch functionalities) attached to it. Desktop apps (meaning apps for computers, including laptops) and web apps both work.

StoryBots Classroom (mentioned previously in this book) was a web app meant to run on computers, tablets, and also on kindergartens' interactive boards. We had to arrange buttons and other interactive components in a way that would work on all of these platforms, which have very different sizes.

If you decided to go with a web app (or a desktop app), consider also the possibility that this will be used on interactive boards and not just on computers and tablets. It's a good opportunity to expand use cases to classrooms and involve educators. Kindergartens especially don't usually have iPads or computers for each child, but there's a good chance they might have a SMART board. So consider carefully (and test!) the placement of each element of your UI, a button that would be very easy to reach with a mouse or by tapping, could be out of reach for a child using an interactive whiteboard if placed too high.

As an alternative, you might consider giving the teacher the possibility of entering in a "SMART board" mode, where the UI changes accordingly, but, just as Jef Raskin once said, I'm not a big fan of "modes."

Think Like a Kid

Every child is an artist. The problem is how to remain an artist after he grows up.

—Pablo Picasso, Spanish painter

One of the most important skills in UX is empathy. The definition of empathy is "the ability to understand and share the feelings of another."[9] It's pretty obvious why this is critical in user experience design. Understanding the people we're trying to help is the first step to find (one of) the right solution(s) to the problem(s) we are aiming to solve.

Empathy can be easy when the people we are designing for are similar to us in gender, age, ethnicity, culture, language, and so on. It becomes harder when we design for users that are different from us. This is also the reason why diversity in teams is very important. Having different perspectives on problems and solutions is always beneficial, also because you often don't even identify a problem if you never experienced it before in your life. If you never spent any time on a wheelchair, it is possible that you'll think it's a good idea, or at least not a big deal, to have ATMs four feet off the ground.

Now, assuming you're a gold medalist in empathy and your team is the most diverse ever, I'm pretty sure no one in the team will be 12 years old or younger (if so, the law enforcement should be knocking at your door any minute now). When trying to come up with a concept for children, it's really hard to put ourselves in our users' shoes, but psychologists have some advice[10] on how we can try going back to our childhood's mental models.

Get Rid of Inhibitions

Wait, keep your clothes on! This is more about loosening up your mental constraints. Author William W. Purkey once said "You've gotta dance like there's nobody watching," and this is what children do. They have no fear of

---

[9]Empathy. (n.d.) In Lexico.com dictionary. Retrieved from https://www.lexico.com/en/definition/empathy.

[10]"How to Think like a Child." *The Independent*. Independent Digital News and Media, 23 Oct. 2011. Web. 31 May 2020. <https://www.independent.co.uk/life-style/health-and-families/features/how-to-think-like-a-child-2116291.html>.

   Himmelman, Peter. "How Thinking Like a Kid Can Make You More Creative." Time. Time, 13 Oct. 2016. Web. 31 May 2020. <https://time.com/4529444/how-thinking-like-a-kid-can-spur-creativity/>.

being judged (when they are really young at least); they dance and sing and play without any concern on how people will see them and what they'll think of them. According to studies, when Jazz musicians improvise, they turn off the area of the brain that triggers inhibitions and self-censoring, and apparently people thrive when they can find some areas in their life where they can really let themselves go.

Have Bad Ideas

This is strictly connected to the previous point. When you let yourself go, you also accept to express ideas that otherwise you would keep for yourself, for fear of being judged.

I highly suggest you keep this in mind during any brainstorming, regardless of it being around a product for kids or something else. Losing this self-censoring and fear of saying the wrong thing or expressing a crazy, seemingly wrong idea can spark other ideas and discussions and really set the ball rolling.

Children don't have this fear, they speak their mind freely, and they don't keep themselves from expressing even the most absurd and wacky of the ideas. They let their imagination go, and you should do the same.

Slow Down and Get Bored

In our everyday life, we lost the ability to get bored, to stand boredom, and just be with ourselves doing nothing. Even when we think or say we're not doing anything, we're actually watching something on TV, or casually scrolling through the endless feed (dark pattern alert!) of our favorite social network, or even considered better for ourselves like reading a book (like this one!), we might not be productive in the strict sense of the word, but our brain is certainly doing something. When we let ourselves wander without a destination and sink in boredom, our imagination kicks in. It doesn't mean you need to stay still on the couch looking at a white wall, you can go for a random walk, with no headphones to separate you from the world, look around, listen to the sounds and the people around you. Let your thoughts flow free of external stimuli that force you to think of something (like a TV show, a book, music, or else), that's what children do and that's why we want to limit screen time for children (see Chapter 7).

Abandon Your Previous Knowledge

For kids many things are new, and this is truer the younger they are. In 1897, Italian poet Giovanni Pascoli published a poem called "Il fanciullino," the little kid, in which he expressed the idea that deep inside each one of us there is still a little kid, able to see the world under a new and unbiased perspective, he can find the truth trough irrational and intuitive thinking, rather than rational reflection. Pascoli used this metaphor to explain the way he wrote his poems, but there is an important message here: try see the world as everything was new to you. As a designer, I can't avoid thinking how this is relevant also in user experience, especially for children. When evaluating an interface, for example, the idea of seeing it as you had no previous knowledge about interfaces can be extremely helpful to establish an empathetic connection with your users.

What to Teach

In Chapter 2, we talked about how a digital product for kids could be or not educational. Again, there's nothing inherently wrong into making an app that has the only purpose of entertaining, as long as this entertainment is not in any way detrimental to kids' physical, mental, and emotional development.

But we also talked about the importance of the educational value, because that's what parents and educators care about the most (remember the "guilt-free screen time" thing). At this point, when you're in your concept phase of your product you might start to question which subject should you focus on, are there some that are more suitable than others? Well, sure, you don't want to teach the physics of fluids to kindergarten's kids, so the first requirement is that you should align with the subjects that are taught in grades compatible with your target audience.

Some products offer a pretty long curriculum spanning from preschool to elementary school, and this means not just that their target audience is much larger but that they can potentially retain users for a long time.

Full Program

One, and probably the most famous, example of this is ABCmouse (ABCmouse.com) by Age of Learning. ABCmouse (Figure 4-2) program is thought for children from 2 years of age to 8, so from preschool to second grade. The program is aligned in subjects and difficulty with what kids learn at school, so it works like a compendium to their ongoing education. Subjects are reading and language arts, math, science, and arts and colors.

ABCmouse offers a full-fledged learning path, covering all the preceding subjects. The path is visually represented as a road with several stations representing lessons and rewards.

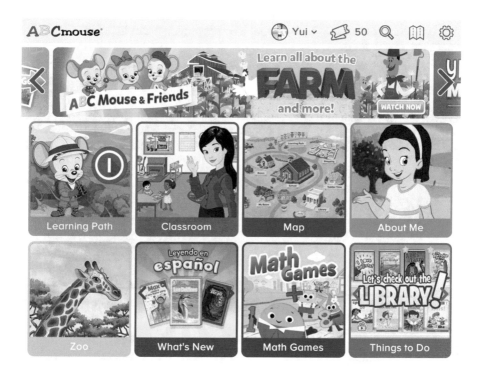

Figure 4-2. ABCmouse welcome screen

Reading

Reading is of course one of the most common skills being taught in apps for children. These apps can offer different approaches and materials; the most common is of course illustrated storybooks. The big difference with printed books is that in digital form books can be interactive and animated and they can offer read-along functionalities, and with speech recognition technologies they can also listen to the kid reading and monitor the progress.

Digital books can provide a higher engagement, thanks to interactive features, and can be customized for each kid with ease. The StoryBots app I've been working on also offered a books section (Figure 4-3). These wonderfully illustrated and animated ebooks offered something special: the child was the main character of the story. With a feature called "Starring You," parents could upload a photo of their child to create their personal avatar, named like the kid; this was then used in the books as the main character, changing the body every time to match the story and the visual style of the book. Plus the books used the kid's real name every time the stories referred to the main character, and all the pronouns were changed according to the kid's gender. All of this plus greatly written stories with amazing looping animations and a masterful sound design made the books very engaging for kids. Today, using more modern technologies like facial 3D mapping, we can imagine even more spectacular products.

Figure 4-3. Books section of StoryBots iOS app on iPad

Language Arts

This subject could be easily combined with reading. Language arts apps can help children learn spelling (complete the sentence or the word, for example, or reorder the letters in a word or the words in a sentence, crosswords, etc.) and new words. Also tracing letters and numbers with their finger could be another simple and fun activity.

An interesting case history is an app called "ABC Gurus" (Figures 4-4 and 4-5), by Colto, an app development agency specialized in products for kids I had the luck to work with on a few projects.

In this app, the kid can pick any letter from the alphabet and decorate it using colors and animated props, then they can trigger an animation that teaches a word starting with that letter.

Figure 4-4. In ABC Gurus, children can customize letters with various props and drawing tools

Figure 4-5. Once they finished customizing, they can enjoy an animation that reinforces the use of the letter

Kids can also take a screenshot of their creation and save it on the device. This app had a tremendous success, due to the simplicity of the concept, the nice illustration style, and the fact that it combines language arts with drawing and creativity. In fact, even if your focus is not a full curriculum, but a specific subject, thinking outside of its boundaries, combining other element, especially if creative, can bring to new engaging concepts.

Math

Math is one of the most common subjects taught in educational apps for kids. Given the variety of topics involved, the relative simplicity in creating exercises, and the importance of the subject itself, math has been the first choice when we created the StoryBots Classroom platform.

Counting, basic operations, basic geometric shapes offer a wide array of exercises to choose from. While language arts require to create each single exercise (you cannot really randomize sentences), math exercise offers a higher degree of automation and randomization. For our platform, we've been able to create a series of basic exercises and then define several algorithms in order to continuously autogenerate random combinations and have a different exercise every time.

SplashLearn offers a curriculum of math activities for children from kindergarten to fifth grade. Their apps are used by 30 million kids worldwide and have been praised by lots of teachers who regularly use the products in their classes. I like how they manage the progression of the activities, going from a more visual and illustrated to more schematic and finally numeric approach, in a way that is very intuitive and ease the kid gently into understanding math concepts that are often a struggle, such as fractions and decimals (Figure 4-6 and 4-7).

What fraction of the pizza slices are left?

2

$\frac{2}{4}$

$\frac{4}{5}$

$\frac{3}{5}$

$\frac{3}{4}$

Level 1

Figure 4-6. A game to learn fractions

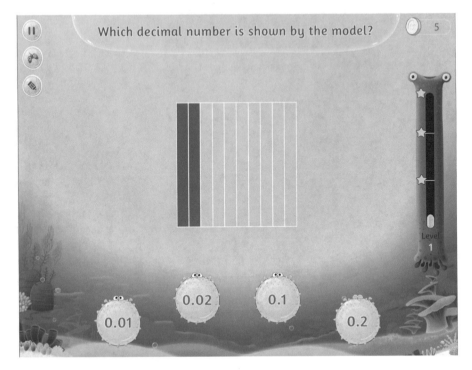

Figure 4-7. A game to learn decimals

Coding

When I was 11, in 1993, I was going to a public school offering an experimental curriculum, including twice the regular amount of English (a second language for me) and coding. For an Italian public school in the early 1990s, teaching kids to code was pretty unusual, and having just received my first PC (a 386 with 2MB of RAM running MS-DOS 6) I was very interested.

The programming language we used was called Logo[11] (Figure 4-8), created in 1967 as an educational tool to teach coding.

---

[11]Logo Foundation. "Logo History." *Logo History*. N.p., n.d. Web. 31 May 2020. <https://el.media.mit.edu/logo-foundation/what_is_logo/history.html>.

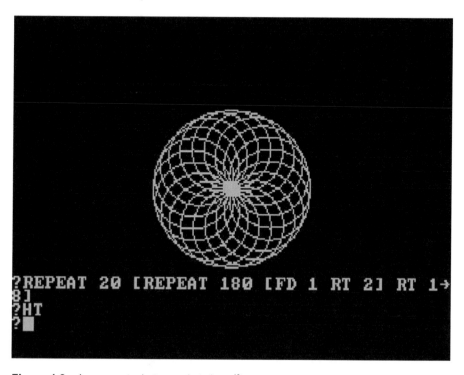

Figure 4-8. A geometric design made in Logo[12]

Despite its age, the methodology behind Logo is still relevant today; in fact, most of the apps in this genre use the same approach, but with contemporary and very compelling graphic styles. For example, *CodeSpark Academy* (Figure 4-9) uses this cell-shaded 3D funny and cute characters, but the logic to instruct them is exactly the same we used in Logo.

Figure 4-9. CodeSpark Academy

Software development is one of the most trending subjects in educational apps for kids. Teaching the basics of how code works can be helpful not just for a future career as developer, but to learn logical thinking and problem-solving, skills that can turn helpful in any moment of our life, regardless of our study or career path.

Hopscotch (Figure 4-10) is a development platform for kids 8–14 years old. The interesting thing about it is that kids don't just create their own games with it, but they can also share their creation with the community of young developers.

Figure 4-10. Hopscotch

There are many products out there to teach children the basics of coding, and apparently the topic is so interesting that even Apple decided to jump in. In 2016, Apple launched *Swift Playgrounds* (Figure 4-11) for iPad, an app to teach, not only their new programming language Swift, but the fundamentals of coding. The app quickly became one of the most downloaded on the App Store in 100 countries, with an average 4/5 rating score. At the beginning of the journey in *Swift Playgrounds*, users are tasked with making a weirdly looking alien reach some red gems through a mazelike floating cubic island. The lines of instructions to make it move forward, left, right, etc. follow the same exact logic we used in Logo to move the "turtle" (that's how the cursor was called) on screen.

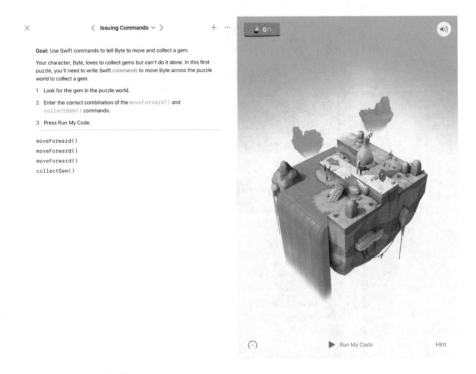

Figure 4-11. Swift Playground

This subject is really interesting because, while all the others can be learned using traditional tools, the digital nature of software development makes it impossible to separate from the use of a device (even though in my coding classes in high school we used to write code on paper first, but we're talking of a prehistoric age).

LEGO is another huge brand that picked up this trend. LEGO Robot Programming for kids is, as the name suggests, a series of products to move the first steps into programming and robotics. There are three lines of product:

- LEGO Boost, for kids aged 7 to 12

- LEGO Mindstorms, from 10 years old and up

- LEGO Education WeDo 2.0, specifically created as a tool for school, designed as a tool for teachers to run engaging classroom projects

The beauty of this product is the combination of creation using hands, so the classic LEGO approach, with the digital building blocks used for coding to bring your creations to life.

Sphero is another great smart device combining a physical "toy" in the form of a spherical robot with a development platform to program it to do obstacle courses and other things. The development happens using logic blocks, like in most of the programming tools aimed to children. The developer can program the robot using data from sensors and using a matrix of LED lights to code a visual output.

The same company previously offered a large catalogue of smart toys, some of which featured widely popular licenses of intellectual properties, such as the now discontinued Star Wars themed droid programming kit.

Another great concept is the one behind *Ozobot*. This is another little robot that works based on a different concept from competitors; besides the classic logic blocks approach, it can also use color blocks to be programmed. The robot has a sensor to read colors, and the developer can use different combinations of color patterns to give it instructions. This means that the robot can work on a digital surface, for example, an iPad, by reading a digital path, or without the use of a device to do the programming, just by using markers on paper. The company promotes this "development without a device" as a differentiating feature, and besides all the considerations about screen time, this possibility represents a great opportunity in the school environment, where having enough tablets for all the students could be a daunting cost (especially for public schools); this way, kids can approach coding using just markers, paper, and a $99 robot. The company is, in fact, exploiting this business opportunity with dedicated bundles for schools.

But why kids should be learning coding? Despite being one of the most paid and sought-after professionals, in 2019 in the United States, only about 10% of STEM (science, technology, engineering, and mathematics) graduates were software engineers. For better or worse, software products are the kind of product that most impacted our lives in recent years. Services like Uber, Airbnb, Facebook, or Tinder changed the way our society is shaped. Getting started in talking the language that will shape the future is better sooner than later, also because there are a lot of ethical implications connected to these products. Explaining how with great power comes great responsibility (thanks, Uncle Ben) will contribute to the rise of a more virtuous generation of startuppers/developers/designers, but, even more importantly though, the future generation will be more conscious of what's behind the screens they'll be using every day, they will have a better grasp of all the implications and consequences of what they do with them. As well as a better understanding of their privacy online, and of the dangers and the opportunities that lie beneath the glass.

Coding is a creative activity. This might seem counterintuitive, as code relies on logic and mathematics, Boolean variables, functions, and many other seemingly uncreative things. But those are just the tools for a very creative outlet. What's more creative than the ability to create something from scratch, just by writing a string of letters (or, for kids, something more visual like logic blocks)? And it's not just about creativity, the fulfilling emotion of seeing your code work (especially after stepping into an annoying bug) is priceless, it can really boost your confidence and self-esteem. With platforms like *Hopscotch*, mentioned earlier, you can share your work and get that ego-boost from the number of downloads or games played.

Other soft skills that coding can teach are persistence, communication, and collaboration. Working on a project often involves asking help from peers, learning from their previous experiences, and sharing ours with the community, in a very healthy balance of competition and collaboration.

You Can Teach Anything

The topics I described in the previous sections are some of the most common choices, but you can actually teach basically anything, from history to geography to languages and more. An interesting finding from a study[13] on app usage for education is that drawing apps can play a positive role in child development, as drawing is an activity that allows children thinking and creating their own prospective of things. Drawing apps can therefore become a supplement to traditional drawing tools (but never a substitute) in contexts where the use of the latter might be difficult or impractical (e.g., travelling).

Finding ways to combine creative outlets to more structured learning, such as the *ABC Gurus* example mentioned earlier, can enhance the experience and make it more enjoyable. The creative activity could even become an incentive; imagine a gamified mechanic in which the child can unlock new drawing tools, colors, stickers, and animations as they solve math operations or other less entertaining tasks.

A couple of examples of unusual subjects: *Star Walk Kids* (Figure 4-12) is an app about astronomy for elementary school children. *Toca Lab: Elements* (Figure 4-13) teaches kids the basics of chemistry and physics, how elements react to changes of temperature, how they change state, and so on.

---

[13]Bozzola, Elena, Giulia Spina, Margherita Ruggiero, Luigi Memo, Rino Agostiniani, Mauro Bozzola, Giovanni Corsello, and Alberto Villani. "Media Devices in Pre-school Children: The Recommendations of the Italian Pediatric Society." *Italian Journal of Pediatrics* 44.1 (2018): n. pag. Print.

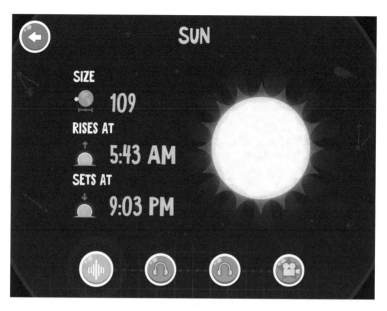

Figure 4-12. Star Walk Kids teaches astronomy to elementary grade children

Figure 4-13. Toca Lab: Elements lets children experiment with funny creatures, who simulate elements' reactions, to learn the basics of chemistry and physics

Passive vs. Active Learning

Teaching and learning can be approached in several different ways. There are a number of teaching methods out there, and they continuously evolve, thanks to new techniques and new technologies being applied and tested.

Defining methodologies to transmit knowledge and educate students is nothing new, it goes as back as the ancient Romans and Greeks and different styles evolved through the centuries and through different cultures, influenced by scientific and social progress.

This is not the book to embark in such analysis, and I'm not the person qualified to do so; in this chapter, we'll talk about the main differences between passive and active learning. We can, in fact, identify these two main groups in which we can divide activities, regardless of the teaching method you might choose.

Why is it important to understand the difference? We extensively talked about the balance between entertainment and fun, and understanding pros and cons of passive and active learning can be a tool, among others, to help us achieve that balance. We also have to consider that each person has a different learning process. There are people who benefit more from reading on their own, others might find practical activities more efficient, others will benefit from audiovisual material, and so on. So understanding the difference between active and passive learning can help us design products that meet the needs of these different styles of learning.

Passive Learning

A student sitting in a classroom listening to a teacher lecturing them, providing notions and facts, is passive learning. When you go to a design conference and listen to speakers on stage is also passive learning, so is watching a documentary or reading a book.

Passive learning is the "classic" approach where someone has the knowledge and passes it to others in a top-down fashion. Since the responsibility for understanding the material falls on the learner, passive learning can teach kids to concentrate and self-manage their learning experience.

Some of the key skills that passive learning helps improve are

- Writing skills
- Listening skills
- Organizational skills

In Chapter 2, we talked about the balance between entertainment and education. There is nothing wrong with a product that focuses on entertainment, for example, an app containing videos and songs, but this content could include educational material, and this would be a perfect way to deliver a passive learning experience. Classics like the ABC song are the perfect example, but there are more contemporary ways of doing children songs, like the super-fun series of rap songs about dinosaurs included in the StoryBots library.

Passive learning doesn't mean boring; with the right writers, musicians, and artists, passive learning can be engaging and fun (see interview with writer Scott Emmons at the end of Chapter 3).

Active Learning

Active learning occurs through discussion and collaboration, critical thinking, problem-solving, in a more "hands-on" approach. Lab experiments are a good example of active learning, so is peer-to-peer discussion where the teacher is just a facilitator. Active learning is more engaging and empowering and generally can be more fun than passive learning, because of the direct experience and the fulfilling possibility of making discoveries on your own.

While passive learning tends to point toward one right answer or solution, active learning facilitates divergent thinking (if you're used to *design thinking* methodology, you should know what I'm talking about) where students explore many possibilities and think more about the big picture than smaller individual problems. This way of thinking helps students in drawing connections between all the elements involved in, and even outside of, the problem scope, the classic "thinking outside the box."

Classic examples of active learning can be

- Hands-on labs
- Group problems
- Peer instruction
- Games and challenges

Which One Is Better?

As you might expect, there is no right answer to this question. Ultimately it all depends on the student and the learning material. A game that is not fun can be more boring and less engaging than a fun music video teaching the same topic.

Depending on the kind of product and resources available, you want to have a mix of both. *ABCmouse* does a pretty good job in including all possible experiences in its learning path; along the way, kids solve puzzles, answer quizzes, watch videos, read stories, and so on. Mixing active and passive learning is the best way to add variety and try to include all learning styles and needs.[14]

Industry Insight: Interview with Dr. Nina Neulight

Nina Neulight is an experienced educator with a master's degree in Learning, Design and Technology from Stanford University and a doctoral degree in Educational Psychology from UCLA, and in-classroom experience having taught grades K through 12. She has contributed to the success and development of educational products for elementary school students that include books, songs, videos, television programs, and a computer-based classroom mathematics program with an at-home component.

Rubens: One thing I point out several times in this book is how having an educational consultant in our team when designing interactive experiences for kids is very important. Mandatory if we want our app to be considered "educational." How do you think is the landscape of educational apps today?

Nina: I'm happy to hear you find it essential to have an educational consultant on your team. Many design teams working on digital education products skip having someone with teaching/learning expertise involved and then often run into problems down the road. Products designed without this type of input

[14]Hahn, Brooke. "How People Learn Best: Active vs Passive Learning." *The Learning Hub.* N.p., 09 Dec. 2018. Web. 31 May 2020. <https://learninghub.openlearning.com/2018/12/10/how-people-learn-best-active-vs-passive-learning/>.

Rodriguez, Brittany. "Active Learning vs. Passive Learning: What's the Best Way to Learn?" *Classcraft Blog.* Classcraft, 20 May 2019. Web. 31 May 2020. <https://www.classcraft.com/blog/features/active-learning-vs-passive-learning/>.

University of Florida. "Adopting Active Learning Approaches." *Adopting Active Learning Approaches - Center for Instructional Technology and Training - University of Florida.* N.p., n.d. Web. 31 May 2020. <https://citt.ufl.edu/resources/student-engagement/adopting-active-learning-approaches/>.

Nisbet, Nigel. "Active vs. Passive: The Science of Learning." *Active vs. Passive: The Science of Learning.* N.p., n.d. Web. 31 May 2020.

Ho, Leon. "Passive Learning vs Active Learning: Which Is More Effective?" *Lifehack.* Lifehack, 28 Nov. 2019. Web. 31 May 2020. <https://www.lifehack.org/858084/passive-learning>.

can lack effective educational context and fail to be truly kid-friendly. We are seeing a serious demand for high-quality, truly educational digital products to use in schools and at home. Schools are more wired than ever before, and more devices are available to students. Parents also want kids to be tech-savvy.

R: What should parents and teachers look out for when deciding which children's product to download?

N: Any product for children should be safe, meaning it has no advertising, violence, or bullying. Overall using the product should be worth the child's time. It's OK to use an app to have fun, but if parents are looking for something that is educational, then it should be just that—educational. Read the product description or have your child tell you about it. I suggest parents try the product themselves and watch their child use it. That will provide parents with a real experience of the product, even just after a few minutes. If you don't like it, don't have your child use it. It's OK to tell your child "no" when there's an app you don't think is right for your child.

R: What are the features a digital product for children should have to be considered educational? What is important to keep in mind when we design one? And what's a mistake to avoid at all costs?

N: Any educational product needs to have learning objectives. Designers of digital products should state these objectives clearly. Your child should be able to tell you in one to two sentences what they will learn after using the product. Many so-called educational games can be played by avoiding any learning. When a child uses the product, parents can ask themselves: Are they practicing something they've already learned or learning something new? Kids need to do both—practice skills and learn new skills and concepts.

The biggest mistake educational technology designers make is to create products that are annoying for the parents, teachers, and other students with loud, irritating noises and graphics. Don't assume kids will be using headphones or will be in a separate room.

R: Being an educational consultant, with a past as a teacher, what do you look for in an interactive experience to use in the classroom?

N: It should be easy to use for both the students and teachers, provide effective learning experiences for all learners regardless of their ability level, and be engaging. It would be great to also have flexibility built-in to the product so that it can be used in different ways—with one student, two students, a small group led by the teacher, or the whole class.

R: How do you advise on testing digital products with kids? What's the best environment? What's the best way to engage with them?

N: Some usability testing is important. Testing should be done early in the design process. The earlier you test, the more likely you are to build a great and effective product. You'll discover issues early on and know if you're on the right path. I like to test with some kids at the target age and some younger and older. Testing should include talking with the kids trying the product to find out what they learned and what they liked most. I find the best way to engage with kids is to have a conversation, keep the test fairly short, and make them as comfortable as possible. Testing intervals at about 15 minutes or less for kids 7 and under is effective since they tend to be pretty wiggly.

R: As an educational consultant and a mom, what's your opinion on screen time? How do you manage it?

N: I think it's important to be aware of how much time kids are using devices as well as how they're using them. There are a lot of great programs out there that give kids experiences they can't have without a screen, and I want kids to get lost in those without feeling stressed that they have a certain number of minutes on screen. As a parent, I never felt a need to set a certain number of minutes my children could have, but some parents might need to set limits. As much as possible, I model good tech behavior, take breaks, and when I have a conversation with a real person, I'm not looking at a screen. Balance is key!

Chapter Recap

- The decision on which kind of devices we want to consider for our product depends on several factors: target (age and skills), business opportunities (ROI), technology, and its pros and cons.

- Consider other devices, besides mobile ones and PCs, for example, interactive whiteboards.

- Thinking like kids is complicated. Try to get rid of inhibitions, have bad ideas, slow down, abandon previous knowledge.

- In educational apps, some subjects are more popular than others. The most sought after are, of course, the ones with more competitors. Try to explore niche ideas.

- Teaching the basics of coding to children can be effectively done with a gamification approach.

- Passive learning and active learning each have pros and cons. Which one is the best is very subjective to each single user. Try to mix the two approaches, when possible.

Gamification

Gamification is one of those buzzwords that started gaining more and more popularity a few years ago. Lots of companies tried to implement some sort of gamification mechanic in their products, but very few did that right.

Why it's important for us to understand what gamification is and how it works? Adding gaming elements to an experience can be a very powerful tool because the idea of play is in our DNA. Humans have played games since forever, as all mammals[1] (and not just them) do. Even if we, sadly, stop playing or drastically reduce its amount, when we grow up, the idea of "game" is very familiar to us and can trigger powerful emotions and memories.

There is a lot of psychology in gamification, as you can imagine, but, to keep things simple, let's keep the conversation on a very practical level, by analyzing which products are effectively using this strategy, what techniques we can apply and how, and, in general, what to do and what to avoid.

If you're designing a learning experience for children, engagement and motivation are two things you absolutely want to achieve. It's easy to have kids motivated when you're designing a pure gaming experience; it is actually hard to do the opposite and avoid them from becoming addicted. But when the scope of your product is an educational one, it's harder to keep your young users' interest and participation. Gamification can help.

---

[1]Sharpe, Lynda. "So You Think You Know Why Animals Play..." *Scientific American Blog Network*. Scientific American, 17 May 2011. Web. 01 June 2020. <https://blogs.scientificamerican.com/guest-blog/so-you-think-you-know-why-animals-play/>.

© Rubens Cantuni 2020
R. Cantuni, *Designing Digital Products for Kids*,
https://doi.org/10.1007/978-1-4842-6287-0_5

But what does "gamification" mean? I think the definition proposed by the *Interaction Design Foundation*[2] is on point:

> "Gamification is a technique where designers insert gameplay elements in non-gaming settings so as to enhance user engagement with a product or service. By weaving suitably fun features such as leaderboards and badges into an existing system, designers tap users' intrinsic motivations so they enjoy using it."

The first thing to notice is that you can't use gamification in games, this is quite obvious. Gamification is bringing gaming mechanics to products that are not games, such as educational products, social media platforms, ratings and reviews platforms, fitness tracking apps, and so on. Too often though, gamification has been implemented as it was a magic potion to boost engagement and content virality. Don't expect a badges system or a digital stickers album to make a boring product engaging and fun. Gamification can help users being more motivated in using the product, but it can't be the only thing you rely on, a common mistake is to think of this technique as a way to make a product a success, but it's not that easy: if a product is bad, gamification alone can't save it.

Nonetheless, if done right, it can be a very successful way to make your product more enjoyable.

Motivation

Motivation is one of the main reasons for implementing gamification into a product. There are two different kinds of motivation: intrinsic and extrinsic.

Intrinsic motivation is when the desire to do something is coming from you; there are no external factors influencing your eagerness to accomplish this thing. For example, you want to learn to play guitar because you want to learn a new skill and play some music.

Being extrinsically motivated, instead, means that the need to do something is influenced by external factors, like getting a reward or avoiding punishment, for example, slowing down while driving to avoid a ticket.

Even if being intrinsically motivated seems the most desirable of the situations, neither of these is actually better—they both work. A study[3] found that praise can help to increase intrinsic motivation in children (and adults), if the positive

---

[2]IDF. "What Is Gamification?" *The Interaction Design Foundation*. N.p., n.d. Web. 01 June 2020.
[3]Henderlong, Jennifer, and Mark R. Lepper. "The Effects of Praise on Children's Intrinsic Motivation: A Review and Synthesis." *Psychological Bulletin* 128.5 (2002): 774-95. Print.

feedback is sincere, promotes autonomy, and conveys realistic and attainable standards. But the same study points out how the effect of praise on motivation is a complex matter, influenced also by the age, gender, and culture of the recipient, and in some cases could even be detrimental.

Gamification comes into play (pun intended) when we talk about extrinsic motivation. It's a tool to provide a sense of reward, being it the top spot on a leaderboard, unlocking new content or any other mechanic that we'll discuss shortly.

In a way, it's a trick to divert intrinsic motivation. A child might not be too intrinsically motivated in learning math, but they might be so in getting the top spot in their classroom's leaderboard at a math game. So gamification, working as an extrinsic motivator, is mediating this intrinsic motivation from math to the game that, at the end of the day, is teaching math to the child.

Some Famous Examples

Let's take a look at some examples of good gamification in digital products. Even though most of these are products for grown-ups, the use they make of gaming components to motivate users is really smart and well implemented. I'm making such examples because it's highly probable you experienced some of these yourself, so it'll be more clear what we mean by gamification, having experienced it firsthand. It'll be also more clear how such techniques are age agnostic and can be used in digital products for users of any age.

Duolingo

Duolingo (Figure 5-1) is an app to learn languages that heavily relies on gamification to motivate its users to make progress and keep on learning.

The app presents a series of achievements and related badges that users unlock by completing activities and by using the app every day to unlock *streak* badges.

Figure 5-1. Duolingo achievements screen

Gamification is so much embedded into Duolingo's user experience that they even did a co-branding activation with *Angry Birds*, the super-popular arcade mobile game by *Rovio*, thanks also to the fact that they both have birds as mascots (we're going to have a full section dedicated to the development and use of characters in digital products in a later chapter).

Nike Run Club

In the fitness world, many products use gamification techniques, because the idea of challenges, rewards, and goals perfectly fits sport activities. One of the first to do so has been Nike with its Nike+ platform (now called Nike Run Club; Figure 5-2). Users can see their results, unlock achievements, compare with and challenge friends, and so on. It's a great tool to boost motivation and keep users engaged helping them developing a daily habit.

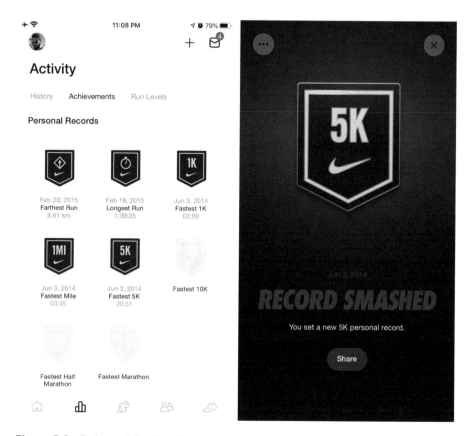

Figure 5-2. Badges in Nike Run Club

Apple Watch's Health Rings

Still in the wellness realm is the example of the health rings in the *Apple Watch*. Apple Watch introduced to the mass public the idea of tracking your daily physical activity to be healthy. The app has three rings (Figure 5-3) to measure how much you move, you exercise, and you stand during your day. By closing these rings, users get notifications and badges to keep them motivated, the more days in a row they close all the rings, the more prestigious badges they unlock. This mechanic is so powerful that some people took as a way of life, arriving to close all rings in a streak years-long.

Figure 5-3. Apple Watch health rings and badges

Swarm

Swarm is the heir of Foursquare (Figure 5-4) which has been maybe the first location-based social media. It introduced the idea of "check-in" as a way for users to communicate with their contacts what restaurant or shop or other business or place they are visiting in real time. The more someone visits places, the more badges unlock; by visiting the same place regularly, a user can become the "mayor" of that place. Even though Foursquare is not as popular as it was when it launched, it's still a good example of gamification applied to real-life events.

Figure 5-4. Badges in Swarm get unlocked by visiting places

Epic Win

Although not as popular as the previous examples, *Epic Win* (Figure 5-5) is, nonetheless, a very compelling case for gamification. Probably one of the most extreme attempts of implementing a gaming mechanic to a utility app, *Epic Win* is an RPG (role-playing game) version of a to-do list app. Doing your tasks and chores lets you earn experience points, unlock loot, upgrade your character, and more.

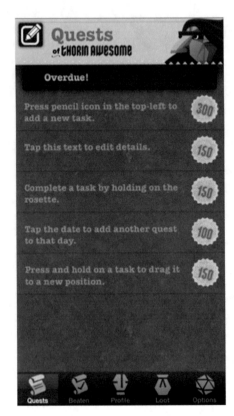 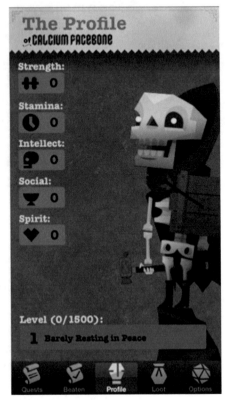

Figure 5-5. By completing tasks, the user can level up their character

There are lots of examples in children's products; here I selected a couple that are interesting because they don't use gamification in the usual educational setting.

One is *Brushing Hero* (Figure 5-6), an app to engage kids into washing their teeth for the right amount of time. Using AR (augmented reality) and face recognition, the kids see themselves as knight fighting monsters. The more they brush, the more they hit the enemies and collect points to level up their character.

Figure 5-6. In Brushing Hero, kids collect points and level up their character while washing their teeth

The second example is the mindfulness app *Stop, Breathe & Think Kids* (Figure 5-7). By completing meditation and mindfulness exercises, children can unlock digital goods. It's a very simple mechanic, but it's interesting seeing it applied to this kind of experience.

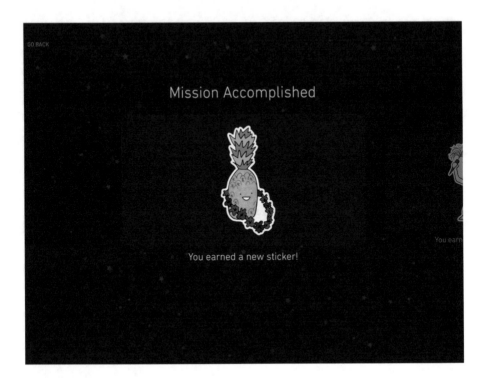

Figure 5-7. By completing missions, children can unlock collectibles

Components

Now that we took a look at some famous applications of gamification, we can identify exactly which are the components and the tools at our disposal to apply it effectively to our product.

These components are what most video games have in common. As I said earlier in this chapter, you can't apply gamification to a video game, simply because gamifying means bringing a non-game product closer to a game, and a video game is, of course, a game already. What you can do is learning from video games. Analyze what makes such products fun and engaging and try to distill those components and, then, experiment how to apply them to a non-game product.

Some examples of gamification components, which can be used alone or in combination, are as follows:

- **Points**: Credits that users can gain through various activities within the product. Points are probably the most classic of the gaming components. They can determine the user's position in a *leaderboard*.

- **Leaderboard**: The list of users arranged by accumulated points. This is another typical element of arcade games, since their early days. The desire to climb up the leaderboard to sit on the top spot is what should motivate users to engage with the product and win points.

- **In-app currency**: Similar to points, but used to trade goods within the platform. In-app credits can be collected in different ways, depending on the product and its mechanics. They are usually a separate thing from point, because they can be spent as virtual money.

- **Experience**: Experience comes from RPGs (role-playing games). In these games, every action and decision we take, every interaction we do with other characters or objects in the game, gives us experience points. Experience basically registers how much we engaged with the game.

- **Level**: This component is an indicator of the experience. It also comes from RPGs, where one of the main scopes for gamers is levelling up their character to become stronger, access new powers, and so on. Levels can unlock special features as a way to motivate users.

- **Progression**: The idea of communicating how far we are into a process can help users to stay engaged and get to the end of it (this is why it's always a good idea to let users know how much of the download or the installation we're going through is done and how much is still ahead of us). Progression is often associated with the level, as it indicates how much experience we still have to get before being upgraded to the upper level.

- **Quests/missions/achievements**: These are the objectives the player has to do in order to get points or experience and/or unlock achievements.

- **Badge**: When a certain goal is reached, a badge is unlocked as a reward. There are normally lots of badges to unlock; they are collected in a specific "trophies" screen, where players can see the ones they got and, usually, the ones they still miss to keep them engaged and motivated.

- **Upgrades**: Upgrades can be unlocked by levelling up or by buying them through in-app currency. They represent some sort of advantage in the game, making it easier to complete missions and unlock badges.

- **Avatar**: In most games (but not all), we do things through a character of some sort. In gamification, this could take the simple form of a profile picture that users choose to identify themselves with other players.

As you can see by reading these descriptions, many of these components are interconnected. Experience points can determine your level that unlocks upgrades to get achievements and unlock badges, and these can reward you with additional experience points and so on. Understanding these mechanics is important to design a proper gamification. Having these ingredients randomly sprinkled on a digital product makes for a very poor (i.e., nonexistent) gamification strategy. It's important, then, to have a good understanding of how you can put these pieces together and create a circular mechanic, like the simple one I described earlier, where the last link of the chain is connected to the first one in an endless loop.

Of course, not all products and their target users are the same. For this reason, it is important to understand your users, what moves them, what are their motivations, but also their skills, in order to develop an effective gamification. The level of complexity you want to achieve depends on this; for example, you don't want to create a complex RPG mechanic, with experience points, levels, power-ups, and in-app currency, for an app aimed to toddlers. In that case, a series of digital collectibles to unlock might be enough. But this probably would not mean too much for older kids, 10 or 12 years old, to whom digital stickers might not be engaging enough. Probably they would be more motivated and entertained by a more complex challenge, with a leaderboard to compare their results with friends and avatars to customize.

Another important thing to understand is that you can't take a finished product and add a layer of gamification over it at the end of the design process. Defining and implementing the gamification strategy is an integral part of the user experience design concept since the very beginning. It's not a module to plug in and out, it's deeply nested into the experience at its very core. When it's just added afterward, you can tell and it's usually a very bad gamification.

Best Practices

Here are some good principles to keep in mind when you design your gamification.

Balance

Gamification is not easy to implement. It's, first of all, a matter of balance between the fun gaming elements and the real purpose of the product. It shouldn't be so overwhelming to transform the nature of your product from its original scope to a full-fledged video game, so balance is very important when designing your gamification strategy. If your app wants to be primarily educational, be sure that's the focus of the experience and the gaming component is the sugar to make swallowing the pill easier.

Go Beyond "BPL"

Badges, Points, and Leaderboard (BPL). That's the most basic (and abused) of the gamification mechanics. It's flexible enough to adapt to a wide array of products, it's true, but you don't want your product to be just another iteration of an old concept, right? Try to go beyond known pattern and already seen experiences.

Tune It for Your Users

Gamification strategies are not a mix and match thing. Each product and its audience requires a custom-made gamification that can't be copied and pasted over to another product. Especially in digital products for children, where a 2-year gap is very significant, as we have seen in the previous chapters, you have to be aware of who are your users, what interests them, what engage them and plan accordingly. The ingredients might be the same, or very similar, but the way you cook them can be very different, when designing for preschoolers or for 5th graders.

Mind the Addictiveness

Yes, we use gamification in order to engage and motivate users, but we don't want them to become addicted to our product. The line between motivation and obsession can be very thin at times, considering the age of our little users we have to be mindful of the time they spend on the device (yes, screen time once again). If we were just looking at raw numbers, addiction would be a sign of a perfect gamification, we maximized the engagement to the point where users find it very hard to stop using the product. Sounds great, right? Well, no. Promoting a conscious and healthy use of our digital product is part of its UX.

Aza Raskin (son of Jef Raskin, the initiator of the Macintosh project at Apple) is largely credited as the inventor of the "endless scrolling" pattern, we are used to experience in Facebook, Twitter, Instagram, and other products.

After a few years this pattern became ubiquitous, he expressed his apologies[4] for creating such addictive interaction, the consequences of which he could not foresee.

As designers, we can't consider engagement a success *per se*; we need to consider it in a larger context where user's well-being is our primary interest.

Chapter Recap

- Gamification can be used to engage kids and motivate them.

- Classic elements of gamification include points, leaderboard, in-app currency, experience, levels, progression, quests, badges, upgrades, avatars.

---

[4]Knowles, Tom. "I'm So Sorry, Says Inventor of Endless Online Scrolling." *News | The Times.* The Times, 27 Apr. 2019. Web. 01 June 2020. <https://www.thetimes.co.uk/article/i-m-so-sorry-says-inventor-of-endless-online-scrolling-9lrv59mdk>.

Safety Measures

It shouldn't surprise anyone that safety must be the number one priority when designing a digital product for children. Safety features for kids' products are not just the obvious ethical choice from designers (and developers and everyone involved) but also a requirement from app stores (both Apple and Google) and from the law (varying from country to country).

Safety means many things, from promoting a healthy use of devices by helping users in keeping their screen time under control, to avoiding bad encounters online, to filtering content not suitable for the age of the user. One of the main weak points in safety of children's products is connectivity.

Connectivity means that our device can access the Internet and therefore an endless stream of possibly harmful experiences for a child.

Even though we can take measures to prevent children to access content outside our product and to connect with strangers, we have to remember that with a simple swipe or tap on the *home button* of the device, they can override any parental gate we can think of.

Now, some Android devices offer a "kids mode" that basically sets the device in a locked mode with functions limited to make selected apps work and nothing else. At the moment I'm writing this, iPadOS and iOS do not offer this feature. They offer a series of parental controls[1] to limit explicit content in music and videos, access to certain web pages, and more, but it's not

---

[1] Apple Inc. "Use Parental Controls on Your Child's IPhone, IPad, and IPod Touch." *Apple Support.* N.p., 24 Mar. 2020. Web. 01 June 2020. <https://support.apple.com/en-us/HT201304>.

R. Cantuni, *Designing Digital Products for Kids*,
https://doi.org/10.1007/978-1-4842-6287-0_6

something you can quickly turn on and off with the press of a button (like the airplane mode) when you want to hand your device over to a child.

Store Requirements

When we're talking about mobile apps, and not browser-based products, both App Store[2] and Google Play Store have a series of requirements in order for an app to be suitable for their children category.

Both platforms have guidelines for developers listing all the requirements they have to follow in order to list their products in the categories for children. In general, what has to be avoided are

- Links to external websites or social media platforms
- Purchasing opportunities of any kind, both virtual or real goods
- Ways to message or communicate in any way with strangers

All the functionalities that could fit into this list have to be hidden behind a *parental gate* (more on this later).

Another important requirement is privacy. Nowadays, we all learned how privacy online is important and should be protected. This is even more important when it comes to kids, so all the digital products that want to find a space on the children category shelf have to be compliant with COPPA[3] (Children's Online Privacy Protection Act) in the United States and GDPR[4] (General Data Protection Regulation) in the EU. Compliancy has to be assured not just at release, but in all future updates of course, and not just that: once the product hit the children category's shelf, even if removed later on and moved to another category, it still has to be compliant, in order to protect all the users who bought it when it was intended as safe for children.

---

[2]Apple Inc. "Building Apps for Kids - App Store." *Apple Developer.* N.p., n.d. Web. 01 June 2020. <https://developer.apple.com/app-store/kids-apps/>.

[3]Federal Trade Commission. "Children's Online Privacy Protection Rule ("COPPA")." *Federal Trade Commission.* N.p., 06 Mar. 2020. Web. 01 June 2020. <https://www.ftc.gov/enforcement/rules/rulemaking-regulatory-reform-proceedings/childrens-online-privacy-protection-rule>.

[4]European Union. "General Data Protection Regulation (GDPR) Compliance Guidelines." *GDPR.eu.* N.p., n.d. Web. 01 June 2020. <https://gdpr.eu/>.

Advertising

Advertising in children's digital products is a very tricky topic. It's a gray area where a big role in what's allowed and what is not is left to the developer's morale, as long as the advertising doesn't conflict with the guidelines described earlier. The content has to be suitable for children, of course, so you can't advertise a gambling website on a product for toddlers (and it wouldn't make sense anyway, even on a purely business perspective), for example.

The problem, until some time ago, was that advertising networks didn't provide a handpicked selection of advertisements 100% safe for children, so relying on these services, made with adults in mind, could be complicated. To solve this issue, we now have advertising networks specifically made for children's products, for example, superawesome.com or kidoz.net. These platforms provide law-compliant content that developers can safely use to generate revenues by displaying advertising or to promote their products across a network of apps, knowing that they'll be targeting their audience.

Advertising and children are a match that is often frowned upon, also because children don't have enough experience to discern what is an ad from what is legitimate content that is part of your product. As adults, having spent years in the digital world, being it apps or websites, we developed what, in UX design, is called *banner blindness*. This phenomenon happens when visitors to a website consciously or unconsciously ignore advertising in the form of banners and similar. It's also why nowadays advertising on website became so aggressive, with news outlets selling their entire background to advertisers and other horrific things like that.

Children are rookies online; they still have to build up that experience made of wrong clicks and taps, pop-ups opening and almost impossible to close, and all the other annoyances that made us (almost) immune. For this reason, my opinion as a parent and as a designer trying to make ethical decisions is that advertising in children's product should be avoided at all times.

Parental Gates

Parental gates are the most common way to prevent kids from accessing sections in your products that are intended for adults, for example, the settings page where parents set up the payment method for a subscription, or a page with links going to external web pages.

A parental gate (Figure 6-1) is a screen, either appearing as a full screen page or a modal, placed in between a section the kids can access and a section they can't, for instance, from the home of the app and its settings.

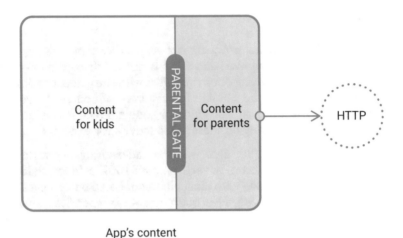

App's content

Figure 6-1. A parental gate locks the access to content made for grown-ups. All accesses to the open Internet are behind this gate

This page (or modal) presents a question that users have to answer in order to get the access to the area dedicated to adults. This question can take different forms, from a simple math operation, to performing a gesture described in a text, or typing a series of numbers written in text on a numeric pad.

In *Eli Explorer* (Figure 6-2) by Colto, the user is asked to perform a gesture on screen.

Figure 6-2. Eli Explorer uses gestures to access the parents' area

Drawing for Kids (Figure 6-3) by Bimi Boo asks the adult to input their year of birth on a numeric pad.

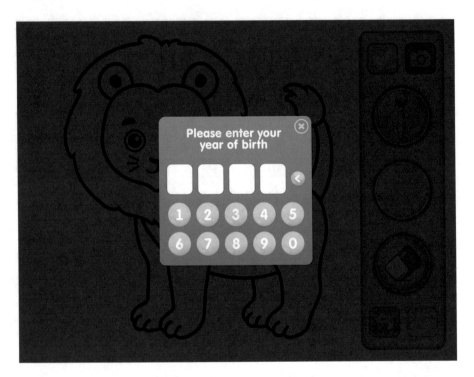

Figure 6-3. Drawing for Kids asks the adult their year of birth

Papumba Academy, as well as many others, asks you to type a sequence of numbers (Figure 6-4).

Figure 6-4. Asking a numeric sequence, from written words to numeric pad

Parental gates can take different forms, as you can see, but what's important to keep in mind is that they have to be easy enough for parents to quickly go through without headaches (you don't want them to solve complicated riddles), but, at the same time, difficult enough for kids to represent a safeguard between them and areas they should not have access to. Another important thing to keep in mind is that these questions or tasks should be randomized. Don't always ask to perform the same gesture or input the same sequence: kids are smart.

This technique for authentication is much harder of course when kids are able to read; in that case, a pin code or a password might represent a better choice. Modern devices also offer biometric authentication based on fingerprints or face recognition, that's a really valid option too.

Social and Messaging Apps

All online social activities, being them messaging or the social networks' kind, require very careful designing when they are meant to be used by children. There are several things to consider and dangers to be aware of. These include

- Encounters with stranger
- Sharing of personal information
- Collection of personal data
- Vanity metrics and their impact on mental health

Some of these things are addressed by app stores policies and government regulations (as we will see in Chapter 10). But others are not, for example, the use of vanity metrics, and it's up to the company developing the app and its designers' ethic to address such issues.

When we design a messaging platform for children, or any form of open communication (e.g., these include comments on content), we absolutely need to be sure that parents and caregivers are in charge of controlling who the children are allowed to get in touch with and to be contacted by. This is the reason why we need parental gates for screens in our product through which kids could access the Web (as discussed in the previous section).

Such apps need to be designed differently for kids and parents, and the latter should have admin powers in order to accept or reject requests to add contacts to the child's account. This is pretty much the promise of all this kind of apps.

It's different when it comes to vanity metrics[5] and their impact on users' mental health. Studies on this subject didn't get to a final answer yet to understand if social networks make us depressed. But it seems undeniable that online social interactions have an impact on our mood.[6]

LEGO Life takes into account many of the issues mentioned at the beginning of this section. The content is moderated by humans, children can't add a photo as their avatar, but they can only choose from a set of LEGO figures, comments are limited to emojis and prewritten responses.

---

[5]Vanity metrics measure popularity of a person (or brand) on social media. Numbers such as *likes* and *followers* are considered vanity metrics. Their use as a way to seek social validation is often associated with depression and self-esteem issues.

[6]Kramer, A. D. I., J. E. Guillory, and J. T. Hancock. "Experimental Evidence of Massive-scale Emotional Contagion through Social Networks." *Proceedings of the National Academy of Sciences* 111.24 (2014): 8788-790. Print.

Where it fails, in my opinion, is in that it provides vanity metrics such as *likes*. Promoting this kind of mechanics among children is an uncertain territory that I, as a parent (and as a designer), would gladly avoid.

The same happens in *PopJam*, a "creative community platform for 7–12-year-olds," as described in their FAQ. Also here content is of course moderated, selfies and photos of people in general are not allowed, unless censored with the stickers provided by the app. The guidelines also explicitly forbid users to post personal information, like real name, address, school name, and so on. *PopJam* too features vanity metrics, such as likes, number of followers, and even a "level" in a classic gamification approach (see the chapter "Gamification") that, in this case, is not used for a greater good such as education, but for mere engagement on the platform.

Sean Herman, founder and CEO of *Kinzoo*, wrote a very interesting book on the subject called *Screen Captured: Helping Families Explore the Digital World in the Age of Manipulation.*[7] You'll find the exchange I had with him on this topic in the "Industry Insight" at the end of Chapter 7.

Chapter Recap

- Safety in kids' products is a priority. All accesses to external links and purchases with real money have to be gated under a parental gate. Access to open communication with strangers has to be avoided at all times.

- COPPA in the United States and GDPR in Europe are regulations that ensure children's data are used properly by developers and their identity is protected.

- A parental gate can use different techniques, from simple math questions to gesture. You can also take advantage of biometric data, such as fingerprint or face recognition.

- Messenger apps should work in a closed network, where only parents or guardians can add contacts for the child.

- Beware of vanity metrics, such as likes and followers in products for children.

---

[7]Herman, Sean. *Screen Captured: Helping Families Explore the Digital World in the Age of Manipulation.* Austin, TX: Lioncrest, 2019. Print.

Interaction Design

Design an Age-Appropriate Experience.

In the previous chapters, we introduced some challenges that designing for children involves in terms of user experience. We talked about the different stages of development of motor skills as these heavily influence the ability to interact with a device and limit or widen our interaction opportunities.

In this chapter, we'll take a deeper look into the interaction challenges (and opportunities) that we'll face when designing a children's digital product.

Children and Adult Users: Key Differences

One thing is clear: children think differently than adults. We will get a little deeper into this topic in the next section, but for the moment, let's just identify some macro-differences between children and adults that affect the interaction and overall user experience.

Kids Develop Very Fast

Both on a physical and on a cognitive level, kids develop quickly. Therefore, a product for a 4-year-old might not be suitable for a 6-year-old; hence, the age groups we have both on the App Store and Google Play Store where apps from children are categorized in intervals of 2 years, "Ages 5 and under," "Ages 6–8," and "Ages 9–11."

Average adults, not affected by any cognitive or physical disability, have more or less the same capabilities regardless of their age. Sure, a person in their 60s might not have the sight of a 20-year-old, but a common banking, or music streaming, or messaging app will perfectly work for all the users despite decades of difference in their age.

Kids Are in for the Journey

As adults, unless we're playing a game, we want our experiences with digital products to be as easy as possible. Imagine using a hotel booking website or buying a flight ticket online. You want things to be easy and intuitive, and you want to find the things you need as fast as possible to complete the task efficiently and without unnecessary headaches.

When kids use a digital product instead, they're in for the journey, not the destination. They don't plan to complete a particular task to get it done as soon as they can; they want to enjoy the experience and find interesting things along the way.

Between the starting situation and the endpoint, kids enjoy what we call micro-conflicts.

This idea is at the core of *gamification*, as we discussed in Chapter 5.

Children Have Less Experience

As we'll see more extensively in the next section, children at these stages of development (from kindergarten to 6th or 7th grade; see the Piaget's four stages of cognitive development later on in this chapter) can't really predict the consequences of their actions ahead of time. They don't have much experience in life, so they tend to be more trusting than adults. Our duty as designers is to make sure our young users will be in a protected and safe environment. This is one of the most important features we have to provide in our products, and we'll inspect what this means in a dedicated section. Regarding interaction with digital products, children have obviously less experience than an average adult. This means that many patterns we take for granted when designing interactions and interfaces for grown-ups are still obscure to children. Creating a mental map of a complex information architecture can be very frustrating, if not impossible, for them. Performing some gestures could represent a double challenge, the first one being to understand them, the second one perform them correctly. The lack of experience with life, but with digital products in particular, makes children particularly vulnerable for what concerns privacy and safety, and exposed to frustration regarding interacting with interfaces.

Kids are not small-scale adults, they are different, and they need different usability guidelines. It's important to notice, though, that many user experience (UX) principles that work for adults and make an experience easy and enjoyable are still valid for children. Consistency is one of these. You don't want to confuse a kid by changing interaction or user interface (UI) component along the way, just like you don't want to do that on adults' products. Simplify the design and avoid cognitive overload is also valid for both kids and adults, and so on.

Table 7-1 summarizes the fundamental similarities and differences.

Table 7-1. Differences and similarities between children's and adults' UX (Data source: Nielson Norman Group)

		Children	Adults
Same	**Following UI conventions**	Preferred	Preferred
	User control	Preferred	Preferred
	First reactions	Quick to judge app/site (and to leave if no good)	Quick to judge app/site (and to leave if no good)
Small difference	**Willingness to wait**	Want instant gratification	Limited patience
	Multiple/redundant navigation	Very confusing	Slightly confusing
	***Back* button**	Used in apps and websites when prominent, but browser *back* button not used (young children) Relied on (older children)	Relied on
	Reading	Not at all (youngest children) Tentative (young children) Scanning (older children)	Scanning
	Readability level	Each user's grade level	8th to 10th grade text for broad consumer audiences
	Scrolling	Avoid (young children) Some (older children)	Some
	Standard gestures on touchscreens (tap, swipe, drag)	Large, simple actions (young kids) Easy and well-liked (older kids)	Easy and well-liked

(continued)

Table 7-1. (*continued*)

		Children	Adults
Big difference	**Goal in visiting websites**	Entertainment	Getting things done
			Communication/ community
	Exploratory behavior	Like to try many options	Stick to main path
		Minesweeping the screen	
	Real-life metaphors e.g., spatial navigation	Very helpful for pre-readers	Often distracting or too clunky for online UI
	Physical limitations	Slow typists	None (unless they have disabilities)
		Poor mouse control	
	Animation and sound	Liked	Usually disliked
	Advertising and promotions	Can't distinguish from real content	Ads avoided (banner blindness);
			promos viewed skeptically
	Disclosing private info	Usually aware of issues: hesitant to enter info	Often recklessly willing to give out personal info
	Age-targeted design	Crucial, with very fine-grained distinctions between age groups	Unimportant for most sites (except to accommodate seniors)

Cognitive Development

We introduced the stages of physical development and the skills associated with each one of them, but there is also another important variable to take into account: cognitive development.

Just like our muscles, also our brain evolves and develops over time when we grow up. The cognitive load that a 12-year-old can manage differs greatly from the cognitive load a 3-year-old can handle. All this has a great impact on the design of our product, affecting, for example, the interactions and the user interface components, not just in their styling but also in the components we can use.

Jean Piaget (1896–1980) was the Swiss psychologist who first recorded the intellectual development and abilities of infants, children, and teens and theorized the four stages of cognitive development in children.

Piaget's interest in the cognitive development of children found inspiration in his observations of his own nephew and daughter. These observations reinforced his maturing hypothesis that kids' brains were not just smaller renditions of grown-up minds.

Up to that moment in history, kids were, to a great extent, treated as smaller versions of adults. Piaget was one of the first to distinguish that how youngsters think is not quite the same as grown-ups do. Based on these observations, Piaget concluded that children are not less intelligent than adults, they think differently. Albert Einstein said about Piaget's discovery "so simple only a genius could have thought of it."

The topic is huge, and it deserves way more pages and expertise than I can provide, but let's try distill the information we need for our design work and define some basic principles and best practices.

Piaget's four stages of cognitive development[1] are

- Sensorimotor (from birth through ages 18–24 months)
- Preoperational (toddlerhood (18–24 months) through early childhood (age 7)).
- Concrete operational (ages 7 to 12)
- Formal operational (adolescence through adulthood)

The ones we're interested in for this book are preoperational and concrete operational. Piaget recognized that a few kids may go through the phases at unexpected ages compared to the midpoints noted earlier and that a few kids may show qualities of over one phase at a given time. However, he was sure that cognitive advancement consistently follows this grouping, stages can't be skipped, and that each stage is defined by new intellectual skills and an increasing comprehension of the world.

Preoperational Stage

During the preoperational phase that goes from 18 months to 7 years old, children are capable of using symbolic thinking, and this could help us, for example, when designing icons for them. Their language becomes more mature and sophisticated, and they also develop memory and imagination. This allows them to understand the difference between past and future and engage in make-believe activities.

---

[1] Cherry, Kendra. "What Are Piaget's Four Stages of Development?" *Verywell Mind.* Verywell Mind, 31 Mar. 2020. Web. 01 June 2020. <https://www.verywellmind.com/piagets-stages-of-cognitive-development-2795457>.

Their thinking, though, is still, for the most part, based on intuition and still not entirely logical. They cannot yet fully grasp more complex concepts such as cause and effect, time, and comparison.

Major characteristics and developmental changes in this phase are as follows:

- Children start to think symbolically and understand how to use words and pictures to represent objects.

- Children in this phase tend to be egocentric; for this reason, they might struggle to see things from someone else's perspective.

- Even though they are getting better with language and thinking skills, they still tend to think in a very practical and tangible manner, with little room for abstract thinking.

Concrete Operational Stage

Elementary-age and preadolescent children, ages 7 to 11, exhibit logical, coherent, concrete thinking.

Kids' reasoning at this age becomes less egocentric and they are progressively more aware of external events. They begin to realize that one's own thoughts and feelings are unique and may not be shared by others or may not even be part of reality.

During this stage, however, most children still can't think abstractly or hypothetically. The egocentric view of the previous stage begins to fade out as children start to have a much better understanding of others' point of view and feelings.

Major characteristics and developmental changes in this phase are as follows:

- In this stage, kids begin to think logically about real events.

- They start to understand the concept of conservation, for example, that an amount of liquid is still the same even if you move it from a thin and tall container to a large and short one.

- Their thinking, despite still very concrete, becomes more logical and organized.

- They begin using inductive logic or reasoning from a specific piece of information to a more general principle.

In both stages though, some key cognitive skills are still immature: the theory of mind, which is the understanding the intentions and emotions of others, empathy is still under construction; cognitive flexibility, meaning the capability of processing conflicting information and switching perspectives; and executive function, the ability to plan and monitor their own behaviors.

Some Design Recommendations

The following are some UX recommendations to keep in mind when designing any kind of digital product for children.

Have a Clear Goal for the Activity

The goal of the activity should be clear, and it should be clear how to achieve it. To do this, you can use different techniques. You can, for example, use an animation to show how to play the game or execute the activity or a simple tutorial with step-by-step instructions.

In this case, don't forget to provide feedback after every attempt at each step.

Feedback is important on any digital product, also for adults, but for kids is mandatory. Providing feedback is not enough though, the feedback must be appropriate for the level of understanding of your users. Earlier we talked about cognitive development and we saw how young kids can't fully grasp others' point of view and feelings. So if your feedback is just based on a character doing different expressions (happy, sad…) according to what action the kid performs, it might not be enough for a 3–4-year-old to understand if the kid is doing something correctly. There could also be a cultural component to consider in feedback based on a character's face expressions and gestures; sometimes they mean different things in different cultures. Consider adding voices with a clear message like "That's right!", "Why don't you try again?", and so on.

Instructions Should Be Designed to the Child's Level of Understanding

This second recommendation is strictly related to the first one. When you explain the activity, be aware of the language you use; it must be easily understood by your users, according to their age capabilities (basically what we previously discussed about cognitive development comes into play here).

Help Kids to Complete Tasks by Using Existing Mental Models and Their Knowledge of the World

As we said before, children can't immediately think symbolically, and even in the older range, we're considering they are still developing this skill. It's crucial, then, to leverage their existing mental models to explain things and guide them through the activity.

A digital art app for adults might use colored circles or squares to indicate the color swatches in a palette (Figure 7-1). For a kid having something resembling a crayon might be better (Figure 7-2). Even though, in truth, my direct experience with my 2-year-old daughter is that after she saw me selecting colors in Procreate, she was immediately able to switch them by herself, calling the palette and then tapping on the color she wanted. But she saw me doing that first, this is the big difference, one thing is learning by observation, another is discovering by exploration.

Figure 7-1. Procreate palette panel

Figure 7-2. Color palette on coloring book from the app Archaeologist

In general, using mental models from a real-life situation they know well can help guide children in completing the tasks. This could let you consider the idea of using skeuomorphic[2] design, and we'll see more about this topic later in Chapter 8, when we'll talk about user interfaces.

Design Self-explanatory Interfaces to Reduce Cognitive Load and Prevent Errors

In the field of UX, by "cognitive load"[3] we mean the amount of mental resources needed to perform a task.

Even though our memory is pretty much infinite, unlike the one of our devices, our *brain power* is not. When we have to deal with an amount of information that exceeds our biological CPU's capabilities, our performances suffer and we take longer to understand information, we miss details, we forget things, and, in the end, we might feel so overwhelmed that we decide to drop the task altogether. As designers, this is obviously something we want to avoid.

---

[2]*Simulating real-life appearance. See Chapter 8 for more on this topic.*
[3]Whitenton, Kathryn. "Minimize Cognitive Load to Maximize Usability." *Nielsen Norman Group*. N.p., 22 Dec. 2013. Web. 01 June 2020. <https://www.nngroup.com/articles/minimize-cognitive-load/>.

The interfaces we design should help users to off-load their *working memory*. Whenever we're engaged into a task, we constantly load our memory with information, and if the interface is not designed properly, we might forget things when moving from one step to the following. This is not because we're distracted or forgetful. We've simply overloaded our working memory with too many information. A well-designed user information should guide us through the task and serve as an aid to our memory, so that we don't need to store all the information.

A common UX myth is the famous "7±2 rule" also known as Miller's law.[4] George Miller was an American psychologist, one of the founders of cognitive psychology. In 1956 he observed how the average person can store in its short-term memory from five to nine items. In UX design this observation is often taken too literally, with designers avoiding designing menus with more than nine choices. The misunderstanding here is that Miller was talking about information that we have to memorize, but menus should be available at all times when using an interface, you don't need to remember all the options by heart. Can you imagine memorizing all the menus and submenus in Adobe Photoshop? Or reducing all its menus to no more than nine items? In both cases, an impossible thing to achieve.

So reducing the load on the working memory means designing experiences that support the user in performing the tasks and achieving their goals, without overloading the cognitive capacity, so that they don't abandon the task and the whole product.

Designing self-explanatory and intuitive interfaces is a best practice for adults' products, but is especially important for those of kids because they are less knowledgeable about user interfaces and digital products in general. The average adult has years of practice using digital products on different devices, so we can now consider some interaction patterns as common knowledge. For example, the scrollbar of a document or web page is now obvious to anyone, but a young child might not understand a scrolling interaction using a similar component. In fact, a lot of components we commonly use in user interfaces don't have anything in common with real-life situations; they are not based on any mental model familiar to kids. So what we consider easy and intuitive as grown-ups will require careful testing to be regarded as the same also for children.

Interactions: Consider Age and Device Peculiarities

In Chapter 4 we talked about children's physical development and how this affects the ability to perform certain interactions or use certain devices.

---

[4]Mcleod, Saul. "Short Term Memory." *Short Term Memory | Simply Psychology.* N.p., 2009.

When we design a digital product for kids, we should have clearly in mind which device is more suitable, not just for the kind of product but for the age group we're designing for.

Use Touchscreens for Younger Kids

Touching an interface is the most natural way of interacting with a device. When we use a keyboard, a mouse, a trackpad, a joypad, or any other input peripheral that lives separately from where the output is displayed, we have a mediated interaction; we act on one device and we get the result on another. When we use a touchscreen instead, the input and output live in the same space. I tap on a button and I can see it reacting under my finger (remember: *feedback*!); I don't rely on a cursor moved by a thing resting on my desk to click on a button displayed on a screen a few inches away from where I move my finger.

It's easy to understand how touch interactions are friendlier for children and how they require less cognitive load. The interface mimicking the physics of real world makes the interaction simple and clear to understand. They are not just simple to perform; they are also intuitive to discover and easy to remember.

In products designed for children, we should limit touch interactions to the most natural and simple ones: tap, swipe, drag.

In my experience, dragging is already a little too advanced for very young kids. In the first version of an educational game I designed, kids moved the character by dragging it around the screen. During tests we noticed how this dragging action required too much finesse in the arm's movement for younger children, so we let the character move automatically where the kid tapped. Older children could still drag it around to enjoy a smoother experience and get a better sense of control over the character, but younger users could easily move it around and enjoy the product as well. It's a good idea to provide, when possible, more than one option to perform actions. Each kid will decide how they want to interact with the product, in the way they find more comfortable.

In products designed for adults, we have witnessed the genesis of a crazy amount of new touch interactions in the past few years. Starting with multi-touch gestures we got to the point where users are supposed to master five-finger pinches, four-finger swipes, up to crazy things like squeezing the phone or using knuckles to knock or draw circles on the screen (these are real examples).

These newer gestures can't be considered "natural" as they are not the digital translation of a real-life action; the reason for their creation is just that we ran out of natural gestures and we needed to come up with new ones to add more functionalities and shortcuts avoiding conflicts (a gesture being associated to more than one result). But these *artificial* (as opposite of *natural*) gestures are more difficult to remember and to master. The lack of a real-life counterpart makes them harder to discover and less memorable; users need to memorize them and remember what we associate to each one of them. Swiping up with three fingers to call the multitasking screen, there's nothing we do in real life that resembles this gesture and its result; therefore, it's something we have to learn and remember how to use when we need it.

The price for these added functionalities is a more complex user experience and more cognitive load for the users.

We can't ask young children to learn and perform gestures outside the ones we classify as natural. And even among them we have to carefully evaluate if they can perform them considering the physical development of the younger users (remember the difference between fine and gross motor skills; see Chapter 4 for more information).

Unintentional Touches

Young children can tap on a touchscreen in a very natural way, but they are not aware of a simple limitation of this technology: if you touch more than one area, the device doesn't know which touch was intentional and which was not. I've observed this happening a lot with my 2-year-old daughter. When holding the screen, they often touch along the edge, so the device registers that as a tap, and, as a consequence, the intentional tapping does not respond, causing frustration and confusion.

We can solve this problem in different ways, each one with pros and cons:

- By implementing a palm-rejection algorithm. Similar to what many digital art apps have, to let users rest the side of their hand on the screen while drawing. This assuming the accidental touch is not with just one finger.

- By defining a safe area around the edges where the app doesn't detect any touch. And this, of course, works only if such unintentional touches are within the safe area.

- By using some sort of timed touch detection. The app rejects touches when held for too long. Also this solution makes sense for unintentional touches caused by how the kid holds the device, not for accidental taps.

I don't consider this a major problem, but when designing touch-based interfaces for kids that are very young, it's something I would consider and eventually test.

Desktop-Based Designs

As we saw in the previous section and in Chapter 4, touchscreens are the best option for younger children. But from the age of 6, we can start considering also computer-based products involving a mouse and a keyboard. In this case though, the interactions that we can use are limited to simple key pressing and clicks on big targets. The size of the target is very important, both with touch interfaces and with mouse-based interfaces, but we'll talk more about this in Chapter 8, which is dedicated to user interface design best practices.

Dragging and scrolling using a mouse or a trackpad requires a level of motor skills that is often too advanced for this age group. The best practice is relying solely on simple keyboard interactions (pressing one key at a time, no combinations of keys) or simple clicks. As mentioned in the previous section, we can still provide dragging interactions, as long as we give the alternative mode based simply on clicking to select the element we want to move and then clicking its destination. Another technique is the *click-n-carry*: the user clicks on the object to drag, this automatically attaches to the cursor, without the need of keeping the mouse button clicked, and then the user can release the object at the destination with another click. Click to attach, move, click to release, instead of click and hold, move, release.

Around age 9, however, kids have developed enough fine motor skills to perform any kind of interaction. Many kids at this age already use gaming consoles and are able to interact with any common device.

Table 7-2 shows a comparison between different age groups and the corresponding suggested interactions and devices.

Table 7-2. Suggested interactions and devices based on age group

Age	Interactions	Devices
2 years old	Single tapping	Tablet
3 to 5 years old	Tapping, swiping, dragging	Tablet (smartphone too, for 5-year-old)
6 to 8 years old	All touch-based gestures Clicking with trackpad, simple keyboard use	Touchscreens and computers with mouse or trackpad
9 to 12 years old	All touch-based gestures Dragging and scrolling with mouse and trackpad, complex coordination between keyboard and mouse	Touchscreens and computers with mouse or trackpad

The use of touchscreens is a good choice for any age, but choosing a touch-based device as our platform is not enough; we have to consider what kind of gestures are suitable for our target age.

Simple natural interactions are the best choice for any age; these include tap, swipe, and, to a certain extent, drag, but this last one should be used carefully and an alternative mode of interaction should always be provided.

We can consider computer-based apps involving the use of a keyboard and a mouse or trackpad on products for kids that are 6 and older. Up to 8 years old though, the only interactions we should consider are key pressing and simple clicks. Dragging, especially on long distances, like from one side of the screen to the other, can be difficult to perform and easily become frustrating. Tests conducted by NNG highlighted how the use of a mouse is preferred by most children, while the same interactions performed on a trackpad are often more difficult for them to perform.

Young Kids Can't Read Yet. How to Solve It?

Ideally younger kids should never be left alone using a device; a parent, a teacher, or a caregiver should be present at all times (see the section about screen time later in this chapter). This seems to solve the problem with the inability of preschool children to read, but it doesn't. Yes, for sure the caring adult can intervene whenever a text requires to be read, but you want the kid to feel in control of the experience and having mysterious symbols requiring someone else for interpretation is not a way to make the child feeling empowered.

Let's see some techniques we can rely on to try solving, or at least easing, this problem.

Use Voice

One way to give young pre-readers a way to navigate interfaces is to use voice user interface instead of relying on text. Voice user interfaces (VUIs) can be used for telling instructions to the kids, but also as an input from the kid to control the product, for example, to make a search, like YouTube Kids does.

On YouTube Kids (Figure 7-3), searches can be either by typing or by voice (tap on the microphone icon first, then talk).

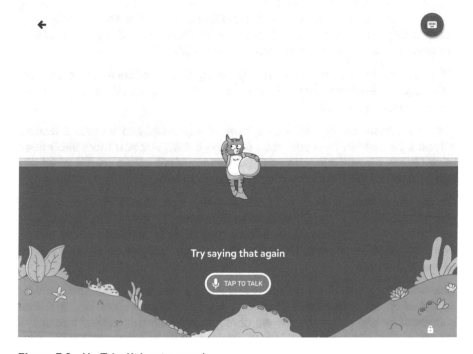

Figure 7-3. YouTube Kids voice search

Once the query is input, the results will show as big thumbnails in a horizontal scroll. Titles are small and secondary; the visual previews of the videos are the main calls-to-action.

Nowadays, voice recognition software is very good and, thanks to machine learning, it's getting better by the day. It's true though that when we deal with toddlers, 2–3 years old, it's often the case they don't speak properly, they mispronounce words or sometimes have a Yoda-like syntax, some of them don't speak at all yet or speak just gibberish. In 2019 during its annual I/O conference, Google presented a voice recognition technology aimed to help people with speech impairments to use *voice user interfaces* (VUIs). They showed a technology capable of understanding nonstandard speech that was really encouraging and frankly close to magic, but something like that can be implemented only by a huge tech company like Google, with access to a never-ending source of data to feed to their machine learning algorithms and continuously perfect the capabilities of the speech recognition software. If they'll release this technology to third parties, smaller companies and startups will be able to introduce more advanced VUI into their products and also products for kids will benefit from this. This is indeed an example of how designing experiences for children has a lot in common with designing accessible products for people with disabilities, so learning to design for very young users will make you a better, more inclusive and more conscious designer also when working on products for adults.

When it comes to using voice as a guide to the experience, there are two main ways of implementing it in your product: using a voice actor or using text-to-speech software.

The first option clearly requires a bigger investment and it's not a flexible solution, because in case you need to change the copy you most likely need to re-record the voice (unless you're lucky enough to have a chance of editing it in postproduction with some cut and paste magic).

On the other hand, a voice actor will ensure a definitely more human feeling to the experience, they will be able to communicate emotions, something a text-to-speech software will not be able to provide (at least at current times), even though this software, much like the speech recognition ones, is getting better and better.

The decision depends on different factors; it's not just about the available budget. For an educational product I was working on, we had activities generated by algorithm so that we could have an almost endless amount of variation; also the assets used were randomly picked every time from a big database of illustrations our artists designed. Given the random nature of how these exercises were generated, we couldn't rely on an actor dubbing hundreds of thousands (probably more) of possibilities; we had to use a software-generated voice.

The decision also depends on the kind of content we're working with. For instructional texts the software option is good enough, but for storytelling, like picture books and such, a voice actor is, of course, the best option.

Use Telling Icons

Earlier in this chapter, we learned how kids' cognitive development is a gradual process and the ability to think abstractly and figuratively is something that happens later in a kid's development (around age 12 and older).

Icons should resort to mental models kids are familiar with. In the software we use every day at work, the classic icon to save a file is often (luckily we started to move away from it) a 3.5 inches floppy disk, something that already in 2007 only 2% of computers still supported. But it's a metaphor that most adults understand. Another example is the classic shape of a phone receiver, even though phones moved away from that design for quite some time now.

With kids we have to carefully think if an icon is using a metaphor that a kid won't be able to understand. For example, we can't use a vinyl record to represent music, as they wouldn't know what that black circle is and what it's been used for.

Use Animation to Guide Through the Experience

Movement gets our attention and can guide us through the experience much more efficiently than static elements. Having an arrow pointing at something could help, but seeing the interaction playing in a "demo" mode is a much more clear.

In *Toddler Games* by Bimi Boo (Figure 7-4), a dragging interaction (which I discourage to use for this age group) is explained with a hand showing how it works at the beginning of the activity.

Figure 7-4. In the app Toddler Games by Bimi Boo, a hand cursor demonstrates the dragging interaction

In this particular case, I think it would have been clearer if we could see the shape moving along with the hand and being placed in its spot.

Animation can be used to show how an interaction works or how to perform a task or play a game, but it can also be used simply to get user's attention on one element at a time and guide them through step by step. The point is using animation can help us to get the attention of the user and distinguish what is an instructional component from the rest. If I use the same element, for example, the hand in Figure 7-4, to point at things on screen and moving it smoothly from one to the other, I naturally guide the user's gaze where I need it to go.

Progressive Disclosure

In user experience design, we talk about *progressive disclosure* when we create an experience that doesn't overload a user with all the information at once, but start with a simple setup and progressively add more information or options or tasks as the user advances in becoming more experienced with the product.

For example, in the example of *Toddler Games* we've seen earlier, the kid is prompted immediately with 3 shapes to drag and 11 empty spots. To make the experience easier to learn without any written instruction, they could have started with one shape and one empty spot, then one shape and two or three empty spots, then two shapes, and so on.

Having a very limited amount of options in the beginning limits the possibility of making errors (and being frustrated by this). With only one possible way of doing something, it becomes much easier to understand the logic and how to progress.

This technique (as well as the others described earlier) is often used in puzzle games, like *Candy Crush* and similar ones, where the progression is very mild, the learning curve is very gentle, and users feel in control, get a sense of reward by completing levels, and are encouraged to progress.

These games are very good at not frustrating their users, and they do it because they monetize over people getting addicted to them. We don't want our users to get addicted to our product, but we surely share with those games the interest in making the experiences as enjoyable as possible and making our users feel in control.

Kid-Friendly Navigation

As adults, our average tech savviness nowadays is quite high. Most people using a digital product that follows the common UX patterns and best practices understand how to navigate through its sections, make a search, login and logout, and so on, without the need of any instruction. We've been trained for decades now; we experienced the evolution of digital products and witnessed the development of user experience design as a discipline. Almost anyone of us, put in front of an average product for the first time, will quickly grasp how it functions and how to complete the most common tasks.

Children still lack this knowledge as they are still new, or quite so, to digital products. Put in front of an interface where affordances are not clearly identifiable, they struggle to understand what is interactive and what is not.

Of course all of these concerns and best practices are more true and valid for younger children, while the older our users are the more they start to have adult-like capabilities. A 12-year-old is able to understand how a drop-down menu works, while a 3-year-old can't.

Affordances

Let's start by defining what we mean by *affordance*.[5] This term has been introduced in user experience design by Don Norman, a researcher, professor, and author in cognitive science and usability engineering, largely credited as the founder of user experience design as a discipline.

In Don Norman's 1988 book, *The Design of Everyday Things*, affordances became defined as perceivable action possibilities. A component's affordances depend on users' physical capabilities and past experiences.

An object (or UI component) that doesn't communicate how it can be operated is lacking affordances.

Take a look at this example (Figure 7-5); the list on the left doesn't provide any affordance to suggest it is a scrollable list and more items are present. The viewport contains exactly eight items and each item is tall one-eighth of the viewport's total height; there is no hint for users to understand they can find more if they scroll up or down.

In the image on the right, items are spaced differently, users can still see there are eight items there, but the last one at the bottom is cut. Moreover, the light gradient on top of the item in the final portion of the list is a visual aid to "trick" users' brain into thinking the list goes on beyond what they can see at the moment. Lastly, the fact that the bottom item is cut and the gradient is at the bottom of the screen suggests the users they can scroll from the bottom up, while if these affordances were at the top of the list, they would have suggested the opposite interaction.

---

[5]IDF. "What Are Affordances?" *The Interaction Design Foundation*. N.p., n.d. Web. 01 June 2020. <https://www.interaction-design.org/literature/topics/affordances>.

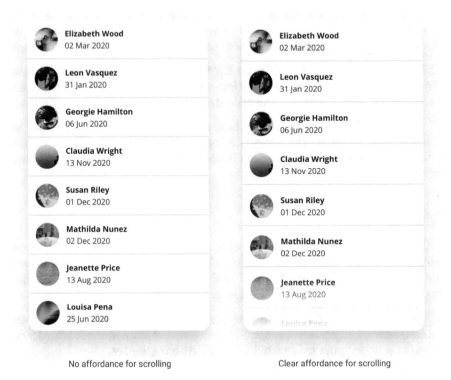

No affordance for scrolling   Clear affordance for scrolling

Figure 7-5. On the right, the fading on the bottom element suggests there is more content discoverable by scrolling up

For children, given their relative inexperience with interacting with digital products, affordances are even more important and need to be even more clear.

There are different ways in which you can make affordances clear for children:

> **Use of color**: Consider using bright colors on interactive elements and more muted colors for backgrounds and noninteractive elements in general (see Chapter 8).

> **Use of animation**: When kids see an animated element, they expect it to be interactive. They are naturally drawn to try tapping on it to see what happens. Static elements are less so, but be aware that animating everything can easily bring too much visual noise and confusion. Use with care.

Adding a shadow to a button is a typical way of suggesting affordance in UI. But for younger children, everything on screen is flat, regardless of any shadow it projects. This metaphor might not provide a good affordance for them.

Moreover, for affordances on kids' products, size matters. We already talked about motor skills and performing interactions; we'll discuss more about sizing and other visual design topics in Chapter 8 on user interface design.

Information Architecture

Simplicity is key when it comes to organizing information in children's product. While adults' apps and websites, depending on their complexity, can have a hyperbranched information architecture (IA), in products for children you want to keep it as flat as possible.

If you have a lot of information and section to arrange, having a flat IA means having too many items in a menu at the same time. It's a trade-off, fewer items in the menu mean a more branched IA, fewer branches in the IA mean more items in the menu, generally speaking.

A hyperbranched architecture can become confusing even for adults; that's why big websites, with lots of information, like a public administration portal, for example, can be very tricky to navigate and require tools like *breadcrumbs*[6] so that users don't get lost. The ideal information architecture for a children's product is made of 2, max 3, levels (Figure 7-6).

---

[6]Breadcrumbs are a tool used in web design to help users understand where they are inside the branches of the information architecture of a website. It's a series of links that starts from the root (the homepage) and follows the path the user took to get to the page they are viewing. It usually looks like something like this: home > category 1 > sub-category 3 > detail page 2 and it's normally located high in the page layout.

LEVEL 0
LEVEL 1
LEVEL 2
LEVEL 3

- Home
  - Games
    - Driving
      - Kitty Town
      - Puppy Ville
    - Puzzles
    - Sports
  - Stories
    - Fantasy
    - Space
      - Planet Party
    - Adventure

Figure 7-6. Simple information architecture

In the IA shown in Figure 7-6, we can see how we have a first level menu with Games and Stories; each of these has a second level, for example, the Games category includes Driving, Puzzles, and Sports, and the Stories category includes Fantasy, Space, and Adventure as second level. The third level is the final one before starting the game or the story. You don't want to go deeper than this, and for children 2–5 years of age, it's best to keep it 2 levels deep or less.

For children inexperienced with the digital environment, it's harder to make a mental mapping of the product, and they can get frustrated if they get lost and don't find what they are looking for. Keep things simple, find the right balance between having too many choices in one screen and going too deep into nesting information.

Menus

Menus can take so many different forms, but not all of these are suitable for children. For example, a drop-down menu requires both motor and cognitive skills beyond young kids' capabilities. As I mentioned in the previous section on information architecture, it's best to try keeping your menus slim, with as few options as possible, such options need to have clear affordances and be clearly separated from the background.

Top navigations and bottom navigations can work for older kids, but for younger users menus should have a central role on the home screen. These options can't be text only; as we discussed in this chapter, younger kids can't read yet, but also older kids might not feel so confident with written words, plus images are way more attractive for any kid. Menu options then are best served with icons or small illustrations that clearly explain what that option is about. When we talked about cognitive development, we said how hard it is for children to think abstractly and metaphorically, so literal representations of things are the best way to go.

On the Emmy-winning *StoryBots* iPad app (Figure 7-7), I worked on as product designer, we went through several iterations of the home screen, before landing on the final one where the menu had a very prominent role, choices were limited, IA was flat, and icons were quite expressive and animated.

Figure 7-7. Home screen of the StoryBots app on iPad

From my observation though, I can tell you that children have a very good visual memory, so even if they don't understand what an option is, and the parent, teacher, or caregiver shows them how it works, they'll memorize and remember also more abstract representations. Of course, this is not ideal; you want your product to be intuitive and clear for kids to be used on their own (not alone, on their own with an adult's supervision is always the best situation).

Feedback

Children want feedback; they need the product to respond to their interactions. First of all, giving feedback is always a good idea, for children and adults. But children have less experience in dealing with devices and digital interfaces, so for them it's even more important to know that whatever they are doing it's having some kind of effect.

This feedback can go from a simple change in the appearance of a component to more articulated animations, including also a sound effect.

Microinteractions

We started this section's conversation talking about feedback, but feedback is just a step of a bigger process called *microinteraction*. What is a microinteraction and why is it important in this context?

If you look online, you'll find a few different definitions of what a microinteraction is. This is mine:

> A microinteraction is the minimum expression an exchange between the user and the system can take.

A single moment of an interaction can be identified as a microinteraction. A microinteraction is never made of a sequence of multiple gestures, as illustrated in Figure 7-8.

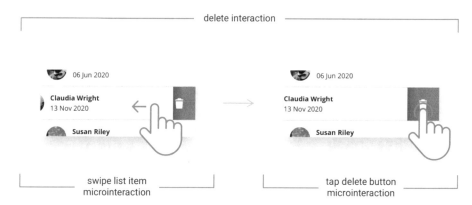

Figure 7-8. Example of microinteractions

A microinteraction can be triggered by the user (e.g., a tap or click on a button) or by the system (e.g., a push notification); this trigger is interpreted according to rules defined in the system and causes a feedback in response; such feedback can be looped or not, depending on the microinteraction, the context, and so on. See Figure 7-9.

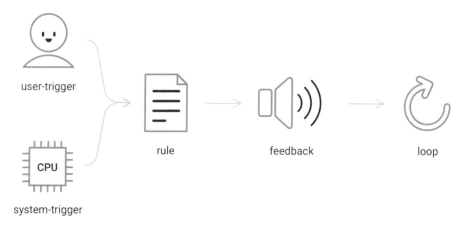

Figure 7-9. The components of a microinteraction

The topic of microinteractions is quite big and here I just want to focus on why they're important in products for children. Microinteractions can help in

- Providing a sense of physicalness to the interface, by bringing metaphors of the real world into the digital space, for example, the reaction of a button when it's pressed (see next section). This helps the transition from real to digital.

- Reassuring the user that what she/he's doing is having an effect on the product.

- Adding some fun to the experience. Remember that children are in for the journey, not really for the destination. When using a digital product, they are experience driven, not task driven as we adults are.

Visual Feedback

Let's see a classic example of feedback provided with a visual design approach: a pressed button.

Buttons are the most common component at our disposal for children's interfaces. For this reason, it's important that these buttons provide the correct experience. A button should always have a *pressed state* (Figure 7-10) to communicate that it's reacting to the child's action. The way you want to communicate this can be different depending on the style of your UI, you want to make the button slightly smaller, as if it was being pushed back, away from the user, or you can tune down the brightness, or make its shadow (if any) disappear, and so on. The important thing is providing a noticeable difference in its appearance in order to make it clear that yes, the app is reacting to my action.

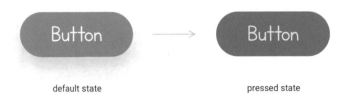

default state   pressed state

Figure 7-10. Change of appearance in a button to communicate a change of state when pressed

The visual feedback you associate to this event could be anything, but it's always a good idea to mimic realistic reactions so that components act in a natural way. A button that expands or rotates when pressed would look quite odd, because that's not the behavior of a physical button.

Animated Feedback

Animation has become more and more important in modern interfaces. We saw in the previous section how different states of a component can provide visual feedback that something is going on. To transition between two states, we can use an animation. An animation enhances the feedback, smoothing the jump between states and inducing a stronger emotional response into the user.

An animated feedback can communicate a message and can make the experience more entertaining. When designing for kids, you can even go a little further compared to when you're designing for adults. While we all appreciate smooth and eye-candy animations in apps, children might find them fun enough to make them sort of a fidgeting device, like a springy reaction of a button or a drawer that overshoots when enters the screen (Figure 7-11), for example. These kinds of cartoony effects work great in interfaces for children.

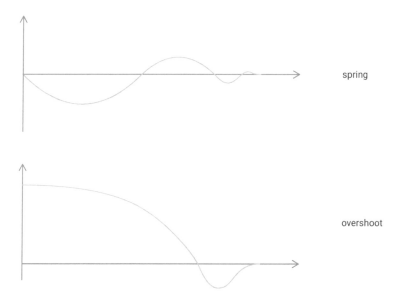

Figure 7-11. Spring and overshoot animation graphs

Animations can really speak to the user. The movement you use should mime the message you want to give. One classic example is a component that shakes left and right to communicate that something wrong happened (e.g., you pressed the wrong button). This movement reminds of someone shaking the head to say "no" (it might not be true for all cultures, so this depends on where the product will be released, who is your target).

Animated feedback can even go further than this. In some cases, you really want to communicate an accomplishment, not just the reaction of a component to an interaction. A correct answer or the unlocking of an achievement (see Chapter 5 on gamification) might deserve some fireworks, figuratively speaking—or literally.

Auditory Feedback

Another very important way of giving feedback is by using sounds. Earlier in this chapter, we analyzed various solutions to help nonreading children, and one of these was the use of voice-overs. That is one perfect example of a sound we can use when kids need a more articulated feedback. But sounds should be also associated to smaller events (i.e., microinteractions), like the press of a button or a swipe or the grab and release of a drag-and-drop interaction.

For these sorts of events, the use of funny and quirky sounds works great along with the kind of animation effects described earlier. Keep it fun, take inspiration from cartoons.

The choice of the sound is very important to complement an animation: it has to match the action, or it will look wrong. A button that pops in and makes a "swoosh" sound would probably look strange, but that same sound on a swipe will fit. The sound should serve as a narration of the animation.

Sounds can be effects or music. In both cases, it's important to make them fun and cheerful, so in the case of sound effects, don't be too realistic and, in the case of music, keep it upbeat. You can use a "sadder" tone when you want to provide a negative feedback for a wrong answer or interaction.

Haptic Feedback

This technology became popular a while ago, and now, on modern devices, haptic feedback reaches incredible levels of realism. Recently, I was talking about this with a friend who did not know nor believe that the trackpad on my MacBook Pro wasn't actually physically moving, but the feedback was just haptic. To convince him I had to turn off the computer and let him feel the trackpad was not clicking at all.

The same goes with the "pop" we feel when using Force Touch on an app on the iPhone. In this case, the animation on the contextual menu that appears helps to deliver the illusion of something actually popping under our fingertip. This example is perfect to understand how the combination of animation and haptic feedback can trick our brain into thinking something physical is happening, while in reality, it's the response of a digital object. If, to all of this, you also add the right sound effect, then you hit the feedback jackpot.

Children need feedback, not just for entertainment but also to understand the product is responsive to their interacting with it.

Visual, motion, sound, and tactile responses combined can be a very powerful cocktail to make your product speak to the user without actually using a voice. They can enhance the user experience greatly, not just by making it more intuitive but also more enjoyable and fun to use. That said, you don't need to combine all of them every time. Sometimes you can use just visual and animated feedback, sometimes you can add a sound, and so on. Crafting the right feedback for an event is a task you shouldn't overlook, because it can really make a difference in a product.

Help Kids Focus

Having kids' attention is not a matter of wanting them to become addicted to our product; this will never be our purpose. We want their attention to teach them how to focus, to help them avoid distractions and learn how to concentrate on a task. The user experience design of our app plays a big role in this.

In our daily life, we are more and more distracted by the many devices around us, demanding our uttermost attention with an endless stream of notifications and stimuli. So much so that big tech companies, like Apple and Google, and all the major smartphone OEMs are moving steps into what we call "digital well-being." Both Android and iOS now offer tools to monitor our apps usage, how many times per hour we pick up our phone, how much time we spend on socials, and so on.

During our work we're also constantly jumping from app to app, from device to device. We answer a message on Slack, we jump into reading an email, we go back to a PDF we just downloaded, we read the notification about an upcoming meeting, and so on. We lost our ability to focus.

The difference is that when we were kids we didn't have such distractions, but kids nowadays potentially do, and this could really hurt their ability to focus. This is why screen time is a very important topic and why we need to help them focus when they are using our product.

But why do kids get distracted in the first place? As I said several times, they are just different from adults and they reason differently. As grown-ups we learned to deal with boring tasks, we know that if something has to be done, it doesn't matter how unpleasant, we have to deal with it. Kids still have to learn that, or they are on their course to do so. If they are presented with a boring activity, they easily diverge toward something more appealing to their natural curiosity. This curiosity is one of the main reasons for their distraction, but it's absolutely a good thing. Being curious is the main driver to knowledge; we don't want to impede that of course, but we want to help them learn how to focus, how to deal with boredom if necessary, and, generally, how to become reliable and dependable.

Avoid Notifications and Pop-ups

In-app notifications on children's product should be avoided, as well as pop-ups that appear from nothing, to promote more games or other distractions.

We have to carefully craft the experience so that users are guided where they want to go without being interrupted by unwanted messages. We pave the road for them to get to their destination and we don't want to put any billboard along the way. Notifications and pop-ups are most of the time annoying for adults; for kids they are not just annoying but confusing and distracting.

Let me clarify one thing though. What's the difference between an *in-app notification* and a *push notification*?

Push notifications are the ones appearing outside of the app sending them. They appear on a system level also when your screen is locked. They're the classic notifications you get when receiving a message or things like that. A developer can decide to send a push notification as a marketing tool to notify users about special promotions or new product releases, and we'll talk about this later when discussing about marketing and other money-related stuff.

An in-app notification is a notification that appears within the app, usually as soon as you launch it. It can be used for different purposes, for example, communicating an update or again to promote related products from the same developer.

Why is this difference important to us? A push notification appears on the device, regardless of the app you are using or even regardless if you're using the device at all; therefore, it's more likely to be something an adult is going to read. More than part of the app experience is part of the device or OS experience.

An in-app notification is an integral part of the app user experience. It's something happening and living within the app, so it's an integral part of the kid's experience when using your product. In most cases, it's also customized using the app design language, rather than the OS' design system.

So, while the push notification speaks to the parents through their devices, the in-app notification speaks more directly to the kids.

Use Animations and Easter Eggs Wisely

We talked about how animation can help guide the user and make the experience more engaging. Animation should not be distracting though, using animation too much, in too many places and without a real purpose, especially during a task (and by task I mean also a game, not necessarily something strictly educational) can drive the user's attention away from where it should be in order to complete it. Same goes with Easter eggs, we all love a bit of surprise in our experiences with digital products. Something like confetti appearing when you write "Happy birthday" to someone in a messaging app, it's a fun, surprising moment that can make the experience delightful and create a more empathic bond with the product. But they should be the exception, not the rule; used sparingly they amaze and entertain, but used too much becomes annoying and, most of all, distracting. Also kids tend to try playing these things over and over, once discovered they become the focus of the experience and that's not what we want.

So, yes please, add some magic moment to your product, but don't overdo.

One Thing at a Time

Multitasking is not for kids. Our pride in the ability to do many things at once should not trick us into trying to teach this skill to kids (and honestly, it's making no good to adults too). To teach kids how to focus is best to prompt them with one task at a time. You can anticipate what's coming next, but you don't want your user to be able to jump frantically from one activity to another.

Break Bigger Tasks into Smaller Chunks

Breaking up a big task into smaller activities is an effective way to help children (and adults for that matter) concentrate and complete it. It's the same principle adopted by ABCmouse in their learning path, which is very long, but organized in lessons, and each one has a series of small activities. They have an overview of the goal and the path to get there, but they'll tackle one thing at a time in smaller tasks.

Another good example is the shopping and cooking activity on *Papumba*.

First, the kid is asked by the grandpa to buy groceries to prepare a dish (Figure 7-12).

Figure 7-12. Step 1, grandpa gives the kid a recipe

Then, at the grocery store, the task is putting on a scale one ingredient at a time (also notice the hand indicating what is the focus of the task; no text involved besides numbers) (Figure 7-13).

Figure 7-13. Step 2, the kid goes to the grocery store and picks the ingredient

When putting ingredients on the scale, kids get immediate feedback from the voice-over, visual effects, and the store clerk character's reaction (Figure 7-14).

Figure 7-14. Step 3, the kid moves the ingredients on a scale to measure the quantity

Once the task is complete, they can see it's marked as done (Figure 7-15).

Figure 7-15. Step 4, when the quantity is correct, a check mark appears next to the item

After buying groceries, they can enjoy feeding grandpa with the dish they helped prepare (Figure 7-16).

Figure 7-16. Step 5, grandpa, kid, and the cat (yes, the cat too) enjoy the recipe

Furthermore, activities like cooking by following a recipe can be defined as *sequencing*. Sequencing means following a series of instructions in a particular order, and it's a kind of exercise that helps improve concentration and focus.

Rewards

Another way to improve concentration is by providing rewards. Rewards are one of the main techniques used in a gamification strategy, as we discussed in Chapter 5.

The idea of a gratification at the end of a task is very powerful, and this works especially well in the form of *blind box gift* and collectibles. This strategy is commonly used in casual arcade games, where players unlock more playable characters the more they play; a very famous example of this idea is the very popular mobile arcade game *Crossy Road* that, as of mid-2020, includes 272 different characters.

The idea of collectible is very often developed in the form of a stickers album. Kids can gain more stickers by completing activities; *Lingokids* (Figures 7-17 and 7-18), an app to learn English, uses this idea.

Figure 7-17. Sticker unlocked in Lingokids

Figure 7-18. Stickers album in Lingokids

Set Breaks

Alternating moments of pure fun to moments of learning can help kids be more concentrated during the latter. If your product offers a learning path with several lessons one after the other, consider having something fun to watch or to do every few learning activities. It could be a fun music video or game, something purely enjoyable just to relax and have a laugh. Knowing something like this is ahead can be also a good motivation to progress further in the curricula.

It could also be a good idea to suggest activities away from the screen or combine the two in a smart way. You could, for example, invite the kid to draw something on paper then snap a photo of the drawing and paint it digitally, add stickers, and so on, or you can take advantage of more advanced technologies such as *augmented reality* (more on this in Chapter 11).

To summarize the ways to help kids focus:

- Avoid distracting events such as notifications and pop-ups.

- Use animations and Easter eggs wisely. Don't overdo.

- Ask the kids to do one thing at a time. Don't overwhelm them.

- Break big tasks into smaller activities.

- Offer rewards.

- Set moments of distraction and recess.

Screen Time, Ethical Approach to Design

The next evolutionary step is into the screen.

—Marc Maron, comedian

When talking about kids' well-being, few topics are more controversial than screen time. It's one of those topics on which every other day a new study comes out stating the opposite of the previous study; it's one of those "Coffee is good for you, no wait, coffee is bad… I mean good, or bad, it depends. But good. No" topics.

Sure, you won't find any study stating that the more screen time, the better, one thing is evident: screen time has to be monitored and accompanied by a parent or caregiver whenever possible. The point is: how much is OK?

As parents, we care for our children's well-being and we want to make informed choices on the time they can spend with a tablet in their little hands, or in front of a computer. But we also know how tempting is the soothing power of the screen to calm down a toddler having a tantrum on a crowded bus or at the pediatrician's waiting room. It's a powerful tool that can't be abused.

As designers, we want to facilitate these informed choices. Nowadays, many digital products (for adults) are designed with the specific purpose of capturing our attention for as long as possible. Social networks are a primary example of these practices. The most infamous example is maybe the "endless scroll" that is now ubiquitous in many digital products out there, not just on social networks. You scroll content with no end in sight, scroll, scroll, scroll, scroll, and fresh content continues to pop up, with no care for your tired thumb. Where is the good old pagination gone?

Digital well-being is now a hot topic in design, both iOS and Android offer tools to track the usage of apps, how many notifications we receive per hour, how often we pick up our phone, and so on. All these metrics have the purpose of helping us to maintain a healthy lifestyle.

With kids, this is even more important.

A fundamental step in kids' development is the imitation of adults, and one of the most common things today's adults do is using mobile devices. Kids get to know these objects very early in their life; according to a study presented during the annual meeting of Pediatric Academic Societies in 2015,[7] most kids have used smartphones and tablets by the age of 2. The study, about kids' exposure to media and electronics, involved 370 parents of children between 6 months and 4 years old.

The survey showed that 97% of the families' homes had TVs, 83% had tablets, 77% had smartphones, and 59% had Internet access. Parents revealed that 52% of kids under the age of 1 year had watched TV, 36% had touched or scrolled a screen, 24% had called someone, 15% used apps, and 12% played video games. The time spent in front of a screen rose proportionally with the age, with 26% of 2-year-olds and 38% of 4-year-olds using devices for at least an hour. To challenge their thinking and problem-solving skills, young children need firsthand experience with actual materials and tools. More than a matter of "time" intended as an amount of minutes, it's a matter of what they are doing during this time. Screen time should be monitored to understand what they are doing during their sessions with a device, which products they're using and how, more than just define a finite number of minutes. Fifteen minutes watching a YouTuber playing pranks are not better than one hour on an app to learn reading and writing.

---

[7]Sifferlin, Alexandra. "Many Kids Under Age 1 Use Devices." *Time*. Time, 25 Apr. 2015. Web. 24 May 2020. <https://time.com/3834978/babies-use-devices/>.

With kids, the correct approach to technology is not just a matter of screen time though. Another important aspect is with whom they spend this amount of time on devices; no substitute for direct interaction with parents or caregivers has been found. It is also true though that children below the age of 3, if provided with the right conditions, can learn words through material on screen, but such conditions imply a parent/caregiver/educator adding additional verbal and nonverbal forms of communication during the video activity. Eye gaze is particularly important in this context. Devices can reinforce what children are learning at school, but unfortunately most of the products on the market are not made for dual use kid+adult, so a challenge and an opportunity is trying to cater for this need.

■ **Challenge/Opportunity** How can we create an educational experience able to engage both the child and the adult, reinforcing their verbal and nonverbal communication?

Björn Jeffery, cofounder of Toca Boca,[8] in an article on his website,[9] stresses the importance of this relationship during activities on devices. He also adds the importance of helping children to choose the right products for them (and make choices for the younger ones), but also observing how these products are used. Are they providing occasions for bonding with friends and siblings in real life? Are they creating moments of sharing and learning? Lastly, Jeffery invites parents to promote variety. Not all apps for kids have to be educational, some could be for fun and entertainment, what is important is variety and balance with other activities outside the screen.

Screen time is more a matter of quality than quantity.

I absolutely agree with his opinion on this. I started my series of articles about designing apps for children (the originator idea for this book) with the same argument: just like pretty much everything else in life, the key is in balance.

What Do Experts Advise?

The American Academy of Pediatrics (AAP) invites parents to avoid screen time for kids younger than 18 months, with the only exception of video-chatting. This activity, in fact, allows a two-way communication between children and relatives, grandparents living far away, for example, that differs completely from passive fruition of video contents or similar activities.

---

[8]Toca Boca is a Swedish app development studio focused on child-friendly applications for tablets and smartphones.
[9]Jeffery, Björn. "In Defense of Screen Time." *Björn Jeffery*. N.p., 08 Sept. 2019. Web. 24 May 2020. <https://www.bjornjeffery.com/2019/09/08/in-defense-of-screen-time/>.

From 2015 to 2018 while I had been living in Los Angeles, my parents were in Italy and my in-laws in Japan. My wife and I became parents of beautiful Yui in 2017, while being very far from both our families. During that time we extensively used FaceTime to let our daughter bond with grandparents on both sides of the world. When she met them in real life, it's been amazing to see how natural their relationship was, even though most of their time together, up to that moment, had been through a screen. So, not all screen time in general should be considered bad; it's important to evaluate what's on the screen, its quality and balance it with all the other experiences happening outside of it.

What else does the AAP suggest? Kids above 18 months of age, as mentioned previously, should never use devices on their own, but always together with a parent or caregiver, for no more than an hour a day. Older children can have longer sessions, but everything has to be balanced by adequate amounts of other activities and sleep (8–12 hours, depending on age); for this reason, they also suggest to keep mobile devices, TVs, and computers out of their bedrooms. It's worth noticing though that this topic is still subjected to continuous studies, and given how digital products are still fairly new in kids' education and development, we'll probably see new results and guidelines as we learn more about the kids–technology relationship in the next few years.

■ **Challenge/Opportunity**   Can we include into our products the right information to educate parents on the right way to make kids approach technology?

Industry Insight: Interview with Sean Herman

Sean Herman is founder and CEO of Kinzoo, a messaging platform for kids and families. He's also the author of the book *Screen Captured: Helping Families Explore the Digital World in the Age of Manipulation*, a book where he explains the difference between good screen time and being screen captured in manipulating endless feeds, where today, even kids seek forms of social validation.

Rubens: *Kinzoo* instantly sparked my curiosity because messaging apps specifically designed for children and their families are not so common. Then I discovered your book, *Screen Captured*, which talks about the dangers of social media (not just for children), the relationship between kids and technology today, screen time, and more. This immediately clicked with me, because my book touches upon many of the same topics—though more on the design side. Your coverage of the need for ethical design for kids caught my attention in particular. In your opinion, what's the most urgent issue we should address in digital products for kids today?

Sean: To me, the most urgent issue in developing digital products for kids is that they need to be designed from the ground up around children's privacy and safety. Kids are not a market that should be expanded into. They have such unique needs, and at younger ages I always argue that privacy and safety are much more coupled than they are when we are older. The range of negative outcomes for children when privacy has been compromised is far greater than it is for adults.

I have a huge concern when existing products simply try to expand into the kids' segment. The reason for this is that adult platforms (especially social ones) are built to optimize two things: growth and engagement. When designing products for children and families, you simply have to trade off some of the growth levers in order to keep it safe, and you shouldn't use persuasive design to drive engagement. Likes, follower counts, and streaks are examples of things that are used to keep us engaged, but I believe there are unintended consequences that come as youth are increasingly relying on getting social validation in these ways.

Retrofitting platforms that were designed for massive growth and scale back to the kids' market is really challenging in my opinion. I think that is why we've seen Facebook Messenger Kids have loopholes that allowed children to connect with strangers in groups, for instance. I don't think it was intentional by any means, but is an example of the challenges.

Even worse, several platforms have simply ignored the fact that kids use them. YouTube is the most obvious example, but we know there are millions of children in North America using Instagram, Snapchat, and TikTok, among others. In my opinion, it was easier for the platforms to ignore that fact, point to their Terms of Use (which say 13+), and simply say the children aren't supposed to be there than it was to comply with COPPA or GDPR-K. But these are also platforms that derive their value from user counts. There hasn't traditionally been enough disincentive from regulators like the FTC to force action, so platforms could enjoy the extra users without changing their onboarding or key workflows to protect the privacy of young users. To their credit, the FTC has done a lot with YouTube to start making some of those changes which starts with acknowledging what we as parents already know— that kids are using those platforms, and in massive numbers.

R: Kinzoo is getting some really great reviews from parents. I'm sure most of them had the same worries you have and I have about digital products for children involving messaging and social activities. How did you address that? And what are you doing to gain parents' trust?

S: First things first, I'm a parent. *Kinzoo* exists because I recognized a gap with connecting my (then) 7-year-old daughter safely and privately with friends and family. I was searching for a solution and could only find Facebook's product, which was a full stop for me. I don't want my daughter exposed to that platform and its business model. We built *Kinzoo* because I think parents need a choice, and those that are more privacy-focused can have an alternative to Facebook. Two things really differentiate us. First, as mentioned, we are built from the ground up around children's privacy and safety. We make decisions as parents first, and as a business second. Second, we have a different business model. *Kinzoo* collects the minimal amount of data possible and will never sell to a third party. We maintain a list of all the data points that we capture from adults and kids on *Kinzoo*. It is a very short list, and we force a justification of how each data point benefits the user, and not *Kinzoo*, in order to continue capturing it. Gender is an example. We'd love to know the split of male and female users on *Kinzoo*, but we couldn't find a way that would help our users, so we made the decision not to collect it for kids.

We will only be successful if we are successful in earning the trust of parents. We make all decisions first and foremost as parents, uncles, and aunties. We will be transparent in everything we do and will always have an open dialogue with those that want to be a part of the community. I'm most excited about that.

R: Kinzoo is kidSAFE certified. In your experience, do certifications like these work toward parents? Do they know what they mean? Is that something they are looking for when deciding which app to download or service to subscribe for their children?

S: Before we even wrote a line of code for the current version of *Kinzoo*, we were compliant with the Children's Online Privacy Protection Act (COPPA), which protects our youngest users when online. In other parts of the world, especially Europe, GDPR and GDPR-K offer similar protections. I think it would be great for all parents to be aware of these regulations, but I'd be willing to bet that less than 5% of parents are, and even less factor it in when considering downloading a new app for their kids.

That's unfortunate, because they are actually very good regulations. COPPA received bipartisan support when it was introduced around the year 2000 after regulators realized that there wasn't much in place to protect the interests of the youngest Internet users. They are very strong and practical regulations that simply require companies to get consent from parents before children can share personal information online and protect users' ability to control their digital footprint. It was relatively simple for us to align with the regulations because they are the sort of things that I want for my own kids online. As a result, there is undoubtedly more friction added to our onboarding process and some of the key workflows like how users connect with each other. But the challenges presented in terms of removing friction are absolutely worthwhile to me because it means a safer online world for children.

I hope we see a day where parents increase their knowledge of the regulations that are in place and make decisions based on them, but I don't think that will happen on a large scale anytime soon. We will continue to try to educate parents on the subject and to be true leaders in the space. In terms of business model, we also know that broadly speaking, privacy is something that virtually everyone says they value, but there is no evidence yet that we are going to be willing to pay for it. So, we know that we have to bring more to the table in order to monetize our product.

R: In an article[10] I wrote in February 2019, I was proposing the removal of Likes on social networks as a way to improve the mental health of their users. By the end of the same year, Instagram decided to remove the like count from its content (I don't take credit for that). You purposefully avoided any mechanic like that in Kinzoo. Do you get any requests from users for such features? Is the lack of these something they notice and appreciate?

S: Likes were never on the table for *Kinzoo*, and we haven't really had any users asking for them. As a messenger, "Likes" are less common anyway, but the mechanism around using the dopamine cycle to increase engagement is something I've educated myself on in great depth. I learned so much about persuasive design that I really got to understand the broader purpose that likes, shares, and follows serve for the platform. It creates a stickiness to the platform by users, but a consequence is that these mechanisms easily become a scorecard for people to measure themselves against and to validate their social standing.

In recent years, youth anxiety and depression have risen dramatically in the United States. I completely disagree with some that are quick to blame the rise in smartphone use among youth as the underlying reason why. It is incredibly silly and irresponsible to paint with that wide of a brush. I believe there are many socioeconomic, political, financial, and social trends that are contributing to these increased levels. In regard to technology though, from my research, the one thing I worry about the most for my own children is them measuring their self-worth in terms of these vanity metrics. I think there is enough anecdotal evidence out there when listening to youth that some, not all, do place a great deal of importance on how many "Likes" their post received rather than focusing on true connection.

I actually argue in my own book that Instagram removed them not to promote better mental health but rather to answer a growing trend of youth deleting many of their posts due to insufficient "Likes." There are many quotes from youth that speak to that fact. Further, if you review the statements made by

---

[10]Cantuni, Rubens. "What If Social Networks Had No "Likes"." Medium. UX Collective, 11 Jan. 2020. Web. 16 July 2020. <https://uxdesign.cc/what-if-social-networks-had-no-likes-e29aedcb1f88>.

those involved with Instagram in regard to removing the feature, they speak to declines in user-generated content due to this pressure as a reason for the change. As a platform, Instagram is only going to be as good as its content, and youth removing huge amounts of their UGC is most definitely a problem for them. Maybe I'm a bit cynical, but I don't think Facebook companies would make a change like that solely for the "greater good." There's always going to be an underlying business reason, which can usually be traced to how they grow and engage their audiences.

I tell the story early in my book of how a child platform that included likes and followers and, even worse, tied rewards to them (like unlocking new features at a certain threshold of followers) was a major catalyst both for my book and for a lot of the design decisions we made in *Kinzoo*. Any platform that anticipates having users under the age of 13 should not include the notion of Likes and Followers. Period.

R: In this book, I point out several times how designing for kids means designing for many different targets: first, we have kids of different ages, then we have parents (and families in general), and often also educators. What is your approach to design both on an interaction and a visual design standpoint?

S: We are trying to accommodate a user base with people aged from 5 to 95, which is incredibly challenging. We partnered with a world-class design company and spent almost a year designing the current version of Kinzoo. The process involved many, many rounds of research, prototyping, and usability testing to get it just right. Or as right as we felt it could be. We took a lot of time and care with each segment—kids, parents, grandparents—to better understand what we had to consider.

From a design perspective, we are dealing with young ages that don't have a lot of learned patterns when it comes to apps, but definitely don't want to be treated as "babies." By that I mean they want room to explore, want to feel some level of ownership and empowerment, but definitely don't want a UI and UX that feels "pre-K." I think YouTube Kids is the classic example of a product that many children reject (when they get to age 5 or 6) because of these reasons.

Parents have a lot of learned patterns and are used to fairly open platforms that make it incredibly easy to connect with others. We have to continue working with parents to give them a user experience that feels familiar but that also educates them on why things have been designed in certain ways to protect the interests of their families. We've really had to do a great job of finding ways to reduce friction so that we can grow, but to educate parents on why we have additional things in place to protect their privacy.

The other aspect that is incredibly challenging is to build a product that is flexible enough to be able to accommodate many definitions of "family." It is relatively easy to design for the nuclear household, but it gets exponentially challenging once we factor in divorces, stepparents and stepsiblings, and countless other permutations of family. We spent an enormous amount of time on this, and it will continue to be something we'll have to focus on improving. We want *Kinzoo* to work for all families, no matter how they define it.

Kinzoo is also unique in that we are a platform shared by both adults and children, as opposed to being two separate platforms like Facebook Messenger and Messenger Kids. As a result, we have many other things we have to consider, but we are excited by the challenge.

R: The COVID-19 pandemic forced families to be apart for quite a long time, with distancing especially affecting grandparents. Did you see a spike in Kinzoo usage during this time? Do you think this event changed the perception of digital products for kids, making us realize we need more specific tools for them?

S: *Kinzoo* was up on the App Store, but for all intents and purposes, we were still running in private beta as COVID-19 began to have a large impact on our lives. In early to mid-March, we absolutely saw a spike at the top of our funnel (website traffic) right to the bottom (downloads and engagement).

I think COVID has many of us rethinking our relationship with technology. I think many parents began seeing screen time as not an enemy, but rather a way to keep their kids connected with friends, family (that might otherwise be very isolated), and classrooms in a very challenging environment. Further, I feel as though parents also got to witness the fact that our kids still often prefer real connection over devices. Many children have outright rejected things like Zoom and would love to get back to the classroom. So, fears about our kids only having friendships through screens should diminish some.

I think *Kinzoo*'s value proposition was never more clear during this period, and I think more parents will think about how to better incorporate technology in the lives of their children as a result. I am hugely optimistic that technology can be a net positive for our kids, but we have to focus on giving them the best of it, without exposure to the worst of it. For us, we focus on connection, creation, and cultivation as our product pillars, which we think is a great start. I hope we see more new companies emerge that put the interests of their users and their privacy ahead of the interests of advertisers. I think with that, we will really being to unlock the vast potential of technology for our children.

Chapter Recap

- Children and adults look for different experiences in digital products.

- Mental models and cognitive load must be age appropriate.

- Children develop in four stages: sensorimotor, preoperational, concrete operational, and formal operational. Each one corresponds to a different level of cognitive and motor skills.

- Interactions should take into account the stage of development of our audience.

- The choice of the device depends also on the age and development of our target. For younger kids, touch-based devices are the best option. Older kids (8 years old and up) can use mouse and keyboard interactions as well.

- Preschoolers can't read. You can help them using different techniques: use of voice-overs, animation, telling icons, and progressive disclosure. These can be combined together.

- Navigation and information architecture have to be simple. Don't let your IA become too branched.

- Avoid distractions like notifications, pop-ups, excessive use of Easter eggs, and animations. Set breaks. One task at a time, don't ask to multitask.

UI Design

How to Create Usable and Beautiful Interfaces for Children.

We made the buttons on the screen look so good you'll want to lick them.

—Steve Jobs, cofounder of Apple

User interface (UI) design is part of the overall user experience (UX) of a product. We often hear or read about UX/UI, as if they were two disciplines going together in parallel, but it's a very misleading way of representing that relationship: UI design is part of UX design, it's a subset.

Just like we mentioned countless times in this book, what's good for adults might not be so for children. In Chapter 7, we talked about how motor skills, ability to perform gestures, reading skills, and cognitive development influence the user experience of a product aimed to children. We have to address all of those concerns when designing the user interface of such a product.

When I was working on the *StoryBots* app for iOS, we tried several iterations, not just for the interaction model. One of the first versions used a rather complex horizontal scroll, crowded with content and call-to-actions, wrongly based on some products for older children. We soon realized that it was way too advanced for our preschoolers' audience and we pivoted to a much more simple design, made of five main sections accessible by five big, bold, animated buttons from the home page.

© Rubens Cantuni 2020
R. Cantuni, *Designing Digital Products for Kids*,
https://doi.org/10.1007/978-1-4842-6287-0_8

On *Eli Explorer* we basically don't have a user interface at all, except for the big bouncing *play* button to start the game and the home button (intended for parents' use). So yes, sometimes the best interface can be no interface. The point is making the experiences as flawless as possible and to do so you often need to remove, rather than add.

Until a few years ago, it was very common to find websites for children designed as if the whole interface was one big picture of an environment, like a room or a street or an entire world. Interactive elements were merging within the background, with a lot of exploration involved. But was this exploration fun or was it frustrating? I'm leaning more toward the second. Nowadays, interfaces have been cleaned up and made more clear, we're more conscious about the hierarchy of components and UX in general. What was the main problem with those "explorative" UIs? Contrast between interactive and noninteractive elements. When the entire screen looks like one picture, you lose most of the affordances that can guide the experience and make it smooth and intuitive. The line between fun and frustrating can be very thin sometimes, and when we design for children we should never forget the basic rules of good design. We start from those and build upon them to make the product suitable and enjoyable for children; we don't scrap all we know about spacing, hierarchy, typography, use of color, and so on.

Exploration is good, children are curious by nature, and we should keep nurturing that, but again: exploration $\neq$ frustration.

It's also important to notice that, while desktop computers' monitors range between 21 and 30 inches (on average) and laptops' are usually 13 or 15 inches, tablets are 10 inches or smaller, and smartphones, in the best-case scenario, have a 6.5 inches screen. So if you're designing a mobile app, consider that it will run on a rather small screen (with the iPad Pro's 12.9 inches as an exception); therefore, it's an excellent idea to keep things simple and clear. For websites it's a different matter, for sure a 20+ inches screen can accommodate more detailed interfaces, but in 2019 mobile accounted for about 53% of global Internet traffic,[1] so we're definitely in a "mobile first" territory, especially with kids (it's more likely parents hand them a mobile device than a laptop).

In this chapter, we'll look at practical ways to make good visual design in our interfaces for children.

---

[1]*"Mobile Vs. Desktop Internet Usage." BroadbandSearch.net. N.p., n.d. Web. 24 July 2020.*

Flat or Skeuomorphic

If you're a rather seasoned designer, I'm quite sure I need not explain you what "skeuomorphic" means, but for all the other people reading and those who need a refresher, let me spend a few lines on the topic.

Skeuomorphism is a term used to describe interfaces in which the components mimic their real-life counterparts. One of the oldest examples is the desktop (intended as the piece of furniture) metaphor that emerged with the first *graphical user interfaces* (GUI). This ancient metaphor is still there today, we use icons, mostly resembling somehow sheets of paper, we put them in folders, literally represented as paper folders, we have a trash bin, and so on. Sure, the graphic capabilities of computers at the time this was implemented for the first time weren't great, so all of those UI components were simplified and not realistic. When Apple released the first iPhone, they felt the need of helping users transitioning to this new device and new input (touch interfaces weren't that popular on consumer electronics). Buttons were 3Dish, the Notes app looked like a real notepad, with paper and leather textures, even app icons were representing realistic objects, unlike their today's very minimal versions (Figure 8-1).

| Skeuomorphic | Semi-flat or Material | Flat |

Figure 8-1. Buttons in different styles and levels of complexity

After some time, when interaction patterns on mobile devices were well established, flat and minimal design became in vogue, also thanks to the popularity of Android and Google Material Design.[2] People didn't need a 3D look anymore to understand a button is a button, UIs became lighter and easier to code and to render, without the need of PNG images for textures. Since iOS 7, released in 2013, also Apple abandoned skeuomorphism and glossy effects (so "Web 2.0"!) and jumped on the flatness bandwagon. On the desktop front, both Mac and Windows are now mostly flat and the same goes for most of modern websites out there.

---

[2]Google. "Material Design Guidelines." *Material Design*. Google, n.d. Web. 23 May 2020. <https://material.io/design/>.

Looking at skeuomorphic interfaces now is like looking at outfits from the 1980s, you get that sense of nostalgia mixed with that "how could we possibly like this?!" feeling.

As I said, there was a reason behind skeuomorphism, and that was exploiting consolidated mental models to make an interface more intuitive, something that nowadays is not necessary anymore, as our mental models are formed directly in digital devices and we don't need the skeuomorphism's training wheels anymore. But what about kids? They're approaching digital products for the first time, the world they have learnt to understand so far is made of real objects, made of tangible materials, in three dimensions.

In my opinion though, I think it's important to make a distinction. I believe we can identify two levels of skeuomorphism: *conceptual skeuomorphism* and *true skeuomorphism*.

By conceptual skeuomorphism I refer to patterns like the desktop metaphor I mentioned before, where we take advantage of established mental models and design a digital rendition of such objects and relative behaviors. Their visual design can be fairly abstract, a file can be a rectangle with a triangle in a corner resembling a fold on a page, it's not really how realistic things look, but how they work.

True skeuomorphism is when the same interaction technique I just described is also rendered with realistic looks, with textures, details, highlights and shadows, and so on to resemble a physical interface.

This second kind of skeuomorphism is more connected to visual trends, and some research[3] has shown how the realism of UI components doesn't positively affect the usability of interfaces, in children on the autistic spectrum. Another study[4] (on adults) even measured that, in many cases, a visually busier design even has a negative effect on the ease of use of a product, affecting especially experienced users, that are slowed down by the complexity and richness of details of the interface components.

A certain degree of true skeuomorphism can be applied if it fits the visual style we decided to pursue (at the time I'm writing this, a so-called *neumorphism* trend is emerging on Dribbble and other design inspiration websites, but no real product has adopted it so far, due to accessibility issues this style creates—surely not good for kids), but don't use it with the intent of making the

---

[3]Shahid, Suleman, Jip Ter Voort, Maarten Somers, and Inti Mansour. "Skeuomorphic, Flat or Material Design." *Proceedings of the 18th International Conference on Human-Computer Interaction with Mobile Devices and Services Adjunct - MobileHCI '16* (2016): n. pag. Print.

[4]Spiliotopoulos, Konstantinos, Maria Rigou, and Spiros Sirmakessis. "A Comparative Study of Skeuomorphic and Flat Design from a UX Perspective." *Multimodal Technologies and Interaction* 2.2 (2018): 31. Print.

interface more self-explanatory and be aware that overdoing it could lead to the opposite result.

Concerns about making buttons (or any other component) tappable can be addressed with other kinds of affordances, such as color, contrast, and animation. You don't need to make a button that looks like it's made of plastic to invite kids to press it; a timed bounce animation on a flat design might be more effective. As a rule of thumb, simpler is better.

Color Palette

When we buy a new car, when we decorate our home, or when we decide what to wear in the morning, colors play a major role in our decisions. It's almost impossible not to be influenced at some level by colors when we are presented with two or more options to choose from. This is also the reason why designers usually make their wireframes in grayscale, to avoid stakeholders (and themselves as well) to be biased toward a particular decision because of colors. Colors can contribute greatly to the success (or the failure) of a product, and each hue is used in design and marketing to send a particular message and inspire specific feelings (blue for trust, orange for playfulness, red for passion, etc.).

On colors and color theory we have an entire history of books. Colors set the mood of your product more than any other visual element; for this reason, the choice of the color palette is one of the most important steps during the UI design process.

A common misconception is that children's interfaces should be very colorful. While it is true that kids, especially younger ones, love bold bright colors, we have to pay attention to not overwhelm them with a color palette that is too generous in hues.

Colors can be one of the tools we use to provide hierarchy in the user interface, we can use colors to distinguish what is in the background and what is an interactive element, for example.

For 2–4-year-olds, color is the primary variable they use to categorize objects. This means that, for a kid in that age range, a big red cube, a medium-sized blue sphere, and a small green pyramid are first and foremost red, blue, and green items, before being a cube, a sphere, and a pyramid or a small, medium, and big object. This is partly true for adults as well, color is a very prominent characteristic of an item, but we're way more capable than young kids in using other means of categorization.

Bold and Bright

In a 1994 study[5] called *Children's Emotional Associations with Colors*, published in *The Journal of Genetic Psychology*, Chris J. Boyatzis and Reenu Varghese tested the emotional reaction of 5–6-year-old children to a series of colors. What they found was that kids tended to express more positive emotion toward bright colors, such as pink, blue, and red, and associated negative feelings to darker, less saturated colors, like gray, black, and brown. This is not really surprising; most of us would probably agree with those children. But while for adults darker colors, like black or dark blue, can also inspire a sense of elegance, luxury, and premiumness, kids don't seem to get much positive vibes from them, even though it's interesting to notice that, according to this study, male children tend to associate positive feelings from darker tones more frequently than females of the same age.

In an older study[6] published in *Journal of Experimental Child Psychology* in 1972, called *Variables in Color Perception of Young Children*, Professor Rosslyn Gaines tested how brightness and saturation of six different hues (green, red, purple, blue, orange, and yellow) effect color discrimination in children. In other words, is it harder for kids to distinguish different colors when these are darker or less saturated (Figure 8-2)? According to this study, yes it is harder. While children are really good at discerning colors when these are bright and saturated, the error rate increases as the colors get darker or more muted. In 1975, this research has been extended to other age groups in another research[7] that tested kindergarteners, fifth graders, high school sophomores, nonartist adults, and professional artists. The error rate followed a similar pattern across all groups, but the frequency of error was linear with respect to age: the younger the group, the higher the error.

---

[5]Boyatzis, Chris J., and Reenu Varghese. "Children's Emotional Associations with Colors." *The Journal of Genetic Psychology* 155.1 (1994): 77-85. Print.

[6]Gaines, Rosslyn. "Variables in Color Perception of Young Children." *Journal of Experimental Child Psychology* 14.2 (1972): 196-218. Print.

[7]Gaines, Rosslyn, and Angela C. Little. "Developmental Color Perception." *Journal of Experimental Child Psychology* 20.3 (1975): 465-86. Print.

Bright and saturated colors | Low saturation | Low brightness

Figure 8-2. Brightness and saturation comparison

■ **Note**   You can see digital versions of the figures in this chapter by going to www.apress. com/9781484262894 and clicking the Download Supplementary Material button.

Finding harder to discern colors when these are darker and more neutral is part of the human biology, but in kids this trait is more prominent. It's a good idea then to keep our colors reasonably saturated and bright, especially for interactive components. For backgrounds and elements that are not interactive, we can opt for more pastel tones (Figure 8-3), to clarify hierarchy and guide the interaction.

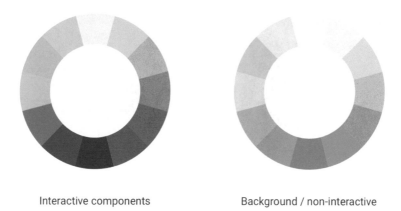

Interactive components | Background / non-interactive

Figure 8-3. Saturated tones for interactive elements, pastel tones for backgrounds

Color Harmonies

Having saturated colors doesn't mean having a full rainbow in all our screens. As I mentioned several times in this book, the general rules for good design still apply on digital products for kids.

Color theory guides us in finding the right harmonies with a series of possible combinations based on the position of hues on the color wheel. We can use these techniques on products for children as well, but we have to keep in mind the limits in color discrimination mentioned in the previous section.

To find color harmonies, we have several strategies that come from centuries of studies in color theory and never changed, simply because the human eye and brain never changed in their physical structure in the last 300,000 years (someone even says 500,000).

Complementary

Complementary colors are hues sitting on opposite sides of the color wheel (Figure 8-4). A primary color and the secondary color derived from the two other primaries are complementary, for example, blue (primary) and orange (secondary color resulting from yellow and red), yellow (primary) and purple (red + blue), or red (primary) and green (blue + yellow).

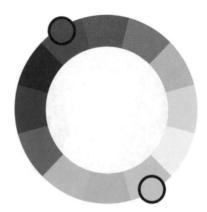

Complementary

Figure 8-4. Complementary colors

With complementary colors you achieve very high contrast, to the point that it can be often too much. It's always a good idea to be cautious when overlapping complementary colors, as they can create a combination that is not very pleasant to the eye, especially with red + green (which is also a very problematic combination for people with deutan color blindness) (Figure 8-5).

Figure 8-5. Be careful with the high contrast with complementary colors

Triadic

The triadic palette is based on three hues equally distant to one another on the color wheel; they are the sitting on the vertexes of an equilateral triangle (Figure 8-6). With triadic harmonies you still have a lot of contrast, but the hues are slightly less clashing than using complementary combinations.

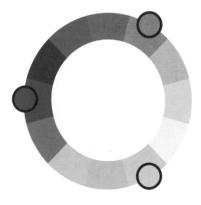

Triadic

Figure 8-6. Triadic colors

Primary colors are triadic, as well as secondary colors, another triadic harmony. To add more options to the palette, while maintaining the triadic scheme, you can work on the value of one (or more) of your triadic colors, making it darker or lighter (Figure 8-7).

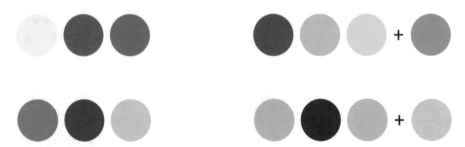

Figure 8-7. Examples of triadic color palettes

Split Complementary

A split complementary color harmony is a combination of the previous two. Start with a color, find its complementary, then move slightly away from it to the left and to the right on the color wheel, to form a narrow triangle (Figure 8-8). The two resulting colors won't be an exact complementary of the initial one, with a contrast that will be less striking and more gentle (Figure 8-9).

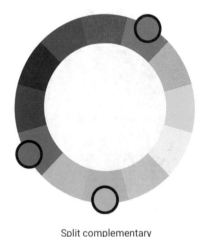

Split complementary

Figure 8-8. Split complementary

Figure 8-9. Examples of split complementary color palettes

Squared and Rectangular

Squared and rectangular harmonies are made of two couples of complementary colors. In the squared one, as the name clearly suggests, the four hues are equally distant from each other on the color wheel, while in the rectangular... well, they form a rectangle (Figure 8-10).

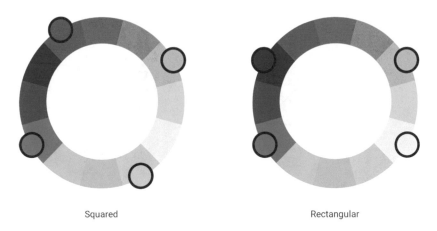

Squared   Rectangular

Figure 8-10. Squared and rectangular

Both squared and rectangular harmonies provide a good balance between warm and cold hues (Figure 8-11).

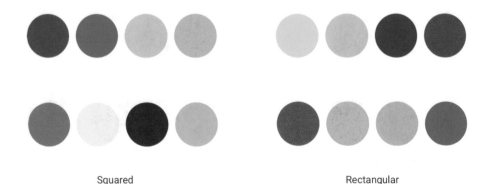

Squared   Rectangular

Figure 8-11. Examples of squared and rectangular color palettes

Analogous

An analogous palette is made by a collection of hues sitting close to each other on the color wheel (Figure 8-12). The result is a more homogeneous set of colors that loses in contrast (Figure 8-13).

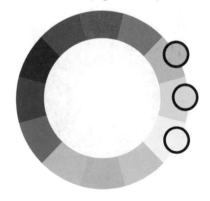

Analogous

Figure 8-12. Analogous

Figure 8-13. Examples of analogous color palettes

Monochromatic

A monochromatic palette is made by just a single color in different values (Figure 8-14). If you think of a black and white photo, that's a monochromatic palette. Besides losing in contrast, compared to other harmonies, a monochromatic palette can be a bit boring, especially for kids (Figure 8-15).

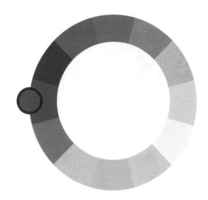

Monochromatic

Figure 8-14. Monochromatic

Figure 8-15. Examples of monochromatic palettes

Suggested Harmonies

We took a look at common techniques to define color harmonies and, considering the information about color perception in children discussed at the beginning of this section, I discourage the adoption of monochromatic and analogous palettes.

The boredom of having limited hues and the lack of contrast make these kinds of harmonies less ideal for children's product. It's impossible to say that any of these methods is absolutely right or wrong, it all depends on the hues you pick, how you use the color palette, the ratio between hues, and more. We can, though, give an indication of which methods (not palettes!) could have a higher chance of success, based on experience and studies on kids (Figure 8-16).

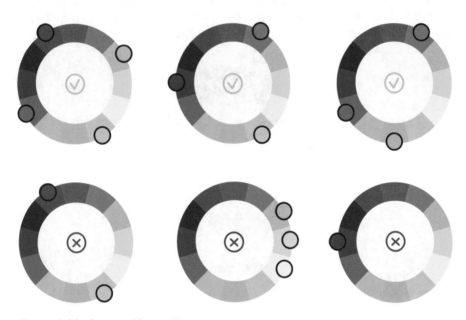

Figure 8-16. Suggested harmonies

Useful Tools

To help you find color combinations, using any of the ratios we just discussed, there are several tools online—for example, http://coolors.co where you decide what method to use to generate the color palette (analogous, triadic, complementary, etc.), play with brightness, saturation and temperature, and more. You can then save or export your palette in different formats, such as a PNG or SVG (Figure 8-17).

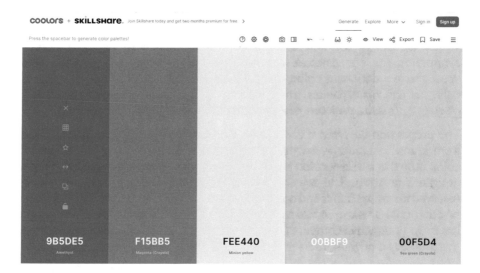

Figure 8-17. Coolors.co

Another one is colorhunt.co, which is a social platform for where designers and artists can create and share color palettes. You can then filter them for the most popular or trendy, the newest, make a search based on a single color you want to include in your palette, and so on (Figure 8-18).

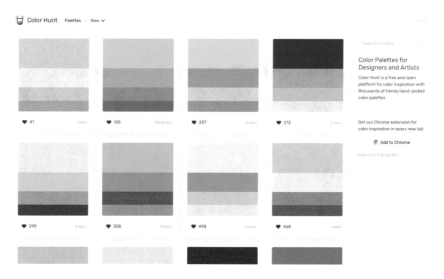

Figure 8-18. Colorhunt.co

Use of Colors for Kids vs. for Adults

When we design for adults, it's a common practice to define a certain proportion for our palette's colors' distribution within our product. We usually have a dominant color, one (or two) auxiliary color(s) and one accent color, as the main palette; then we can have a series of colors used less frequently (maybe on icons or illustrations, on warning messages, etc.).

The proportion between these can be different from product to product and from brand to brand. A frequent distribution is 6:3:1 (60% for the dominant color, 30% the auxiliary color, and 10% the accent), which is the golden ratio (a designer's favorite). The bigger portions (dominant and auxiliary) are usually reserved for backgrounds and are usually more neutral colors (black, white, grays, or other cold or warm muted tones), while the accent is for where the attention has to go (interactive components, such as buttons, switches, check boxes, etc.), this is usually a brighter, more saturated color (blue, purple, yellow, etc.).

In the example in Figure 8-19, we can see how the blue accent color is used in the bottom navigation to indicate which section of the app is currently displayed, and on the *calls-to-action* (the buttons under each piece of content). Generally, we use just one accent color, and that's the same every screen of the app. It's easier for the users to understand at a glance what is important on that screen. Designers use the accent color to guide the experience, to help users find their way through flows, and, sometimes, to suggest one action instead of another ("Subscribe" rather than "Cancel").

The accent color is also part of the product's or company's brand, so it serves also communication purposes, not just interaction ones.

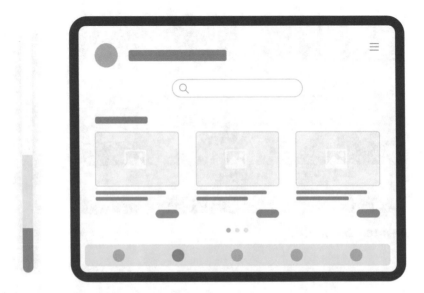

Figure 8-19. Use of colors on an app for adults

In products for kids, we can use more than one accent color and we can use this variety to our advantage. As previously said, children, especially the younger ones, use color as the primary variable to distinguish among different objects. For this reason, we can, for example, use a different color to identify each section of our app or website. Purple is videos, blue is games, yellow is storybooks, and so on (Figure 8-20). Each one will be a different icon, but the first thing they'll notice will be the color, and they'll learn quickly that to go watch a video they'll need to tap on the purple item on screen.

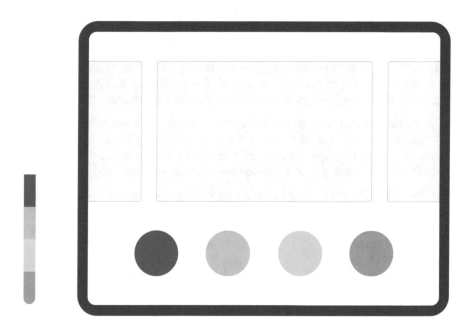

Figure 8-20. Use of colors on an app for children

Colors won't just make our app or website more colorful and playful, but they'll serve a specific UX purpose, making navigation more memorable.

There is one thing to pay attention to, though. While using different colors to identify sections can help children find their way within our digital product, having more than one color on different instances of the same component can be confusing. An example: an arrow to go back to the previous screen should always have the same color (and design, and feedback).

Variety in colors is good when it helps users understand where they are or where they want to go. Or to identify one action as opposed to another; for example, "blue is to create something" and "red is to delete something," but then any button that creates something will have to be blue and any button that deletes something will need to be red. Consistency is of uttermost importance.

The bottom line is colors can be an additional aid to enhance children's user experience, helping navigation and flows. You can assign different colors to different sections and actions, but then you have to keep them consistent throughout your product.

Gender Bias in Colors for Children

We learn that pink is for females and blue is for males since a very young age (Figure 8-21). In Italy we use to place a ribbon on the front door of the house of newborn's parents; such ribbon is pink if the baby is a female and light blue if a male. When and where this norm originated from is still up for debate.[8]

Jo B. Paoletti, a historian at the University of Maryland, wrote a book on the subject[9] titled *Pink and Blue: Telling the Boys from the Girls in America*. In her book she goes through decades of baby's items and fashion trends to try identifying when this popular gender-based color relation originated.

Her belief is that up until the Second World War there wasn't a clear direction to the point that the June 1918 issue of the *Earnshaw's Infants' Department*, a trade magazine for baby clothes manufacturers, said: "There has been a great diversity of opinion on this subject, but the generally accepted rule is pink for the boy and blue for the girl. The reason is that pink being a more decided and stronger color, is more suitable for the boy; while blue, which is more delicate and dainty is prettier for the girl," so completely subverting the current belief. This quote though gave birth to the urban legend that, at that time, the rule was indeed pink is for boys and blue is for girls, while in reality that was probably just one of the many suggestions or attempts to find a common ground for marketing purposes, far from being the norm.

Several studies tried to investigate if the reason for females preferring pink might be hardwired in human genes, but no evidence of such thing has been found, and the preference seems to have exclusively cultural roots. The fact that this seems to be so globally widespread and crosses the border of many different modern cultures is just because of the popularity of Western/American pop culture, as shown by a 2018 study[10] done on children from Hong Kong.

---

[8]Wolchover, Natalie. "Why Is Pink for Girls and Blue for Boys?" *LiveScience*. Purch, 01 Aug. 2012. Web. 23 May 2020.

[9]Paoletti, Jo B. *Pink and Blue: Telling the Boys from the Girls in America*. N.p.: Indiana UP, 2012. Print.

[10]Yeung, Sui Ping, and Wang Ivy Wong. "Gender Labels on Gender-Neutral Colors: Do They Affect Children's Color Preferences and Play Performance?" *Sex Roles* 79.5-6 (2018): 260-72. Print.

Figure 8-21. On the left: The Pink Project - Jeeyoo and Her Pink Things, Seoul, South Korea, Light jet Print, 2007 ©JeongMee Yoon. On the right: The Blue Project - Seunghyuk and His Blue Things, Gyeonggi-do, South Korea, Light jet Print, 2007 ©JeongMee Yoon

The only certain thing is that the gender association we know today started to gain popularity in the United States since the 1950s, getting real momentum in the 1980s, when advertising, marketing, and consumerism in children's products (and in general) hit an all-time high. Toys like Mattel's Barbie contributed to make pink the epitome of the kind of femininity we're questioning today. In fact, Mattel is questioning this as well. In recent years, Barbie has been rebranded as a gender-gap opposer, with the "You Can Be Anything" campaign.

With our daughter, my wife and I try to avoid pink clothing and toys because we don't want her to be influenced by what society says girls should like and do. We don't push any gender-specific toy or activity, we tried to foster her curiosity toward many different things, leaving to her to decide if she is more interested in animals or vehicles or music or cooking or else, solely based on her real interests and tastes (that, as it is normal for a toddler, change very frequently). We're also very careful about the movies and TV shows she watches, for the same reasons. Nonetheless, around age 2, she started to be inexplicably attracted by pink. If put in front of several color options, one of which being pink, she picks this most of the times. That might be just her personal taste, just a random preference, it could have been green, but it's pink, or it could be the influence of external factors out of our control, most probably other young girls.

What about digital products? Should they follow the same gender-defined pattern? I don't think so, for a couple of reasons.

First of all, we should start leaving this old-school marketing heritage behind. In a world that is craving for inclusivity and equality, perpetuating the female=delicate=pink and male=strong=blue seems pretty backward.

Another reason is that if your product is not tied to a very gender-targeted brand (like Barbie), it's a good idea to be as neutral as possible, regardless of the activity. Toca Boca's digital toys are perfectly enjoyable by males too, even when the subject is something that was traditionally associated with females, like hairdressing or dressing dolls and characters. This of course opens up your product to target a wider audience.

Icons

Icons are probably the most common visual aid we have in user interface design. They help the user to identify functions without the need to read text, or they are complementary to a text label to clarify a concept. We can find icons on toolbars, in buttons (alone or together with a text), in any kind of menu, in navigation components (e.g., bottom navigation bar in mobile apps), and so on.

In 1973, researchers at Xerox PARC developed *Alto*, the first personal computer featuring a GUI (graphical user interface; the term UI we commonly use today, technically, could also refer to an interface made of physical buttons and knobs, but, to speak today's common language in the industry, I'll refer to GUI as UI). This revolutionary experiment proposed for the first time the *desktop metaphor*[11] we still use on our computers today, and this representation introduced the idea of the *icons*, tiny drawings depicting a concept.

The use of icons has become even more important on mobile devices, where each app is launched from an icon that serves as the primary way to recognize an app from the other at a glance. Moreover, mobile apps can count on a smaller real estate for their UI, compared to computers, and icons can help in using such space as efficiently as possible.

Abstraction in Icons

The design of an icon can be more or less literal, depending on the concept it needs to represent, but even the most abstract ones, once they become the standard, are usually easy to recognize. See the examples in Figure 8-22.

---

[11]The desktop metaphor is the use of real-life mental models, derived from a classic office scenario, to represent digital components. For example, pieces of information are called "files" and are organized into "folders." The home screen, where we open folders and files, is called "desktop." We delete files by throwing them in a "trash bin," and so on.

Figure 8-22. Common abstract icons. Most users nowadays understand their meaning

One of the most discussed icons in recent time by UI professionals is the so-called hamburger menu. Hated by many designers, yet very popular in slightly different shapes and forms. It's not totally correct to say that the hamburger icon is abstract; its design represents the items listed in a menu. Nevertheless, the high level of synthesis used for this design makes it quite hard to say that this is a literal icon. Thanks to its popularity in apps and website though, most users today know what it means and know what to expect when tapping or clicking on it.

Even icons that were very literal when initially used can become the abstraction of a concept with time. Some of the icons we use on adults' products refer to old technologies that don't have much sense today, but their visual representation has become a synonym for the action they perform.

A couple of classic examples: save and call (Figure 8-23).

Figure 8-23. "Save" icon and "call" icon

The floppy disk icon to represent the save action was very literal when, decades ago, we used to save data on that kind of physical storage. The same goes with the icon for making phone calls. The only way to find a receiver like that on a phone is by going to an antiques store. I'm pretty sure the majority of Gen-Zers never saw either of those things in real life, not to mention used one. Yet, the icon to make FaceTime calls in iPadOS (released in September 2019) is that kind of phone receiver, and the same design is ubiquitous on the vast majority of smartphones on the market today.

Lately, the save icon is getting updated with a different one: an arrow pointing down toward an open rectangle or a "tray" of some sort (Figure 8-24). But the same visual is often used to indicate "download." The two actions can sometime overlap, but they are not exactly the same. The floppy disk is still very popular.

old save   new save

Figure 8-24. Old save icon and new save icon

We saw, in Chapter 7, how children's thinking is very concrete until formal operational phase of cognitive development (around age 12 and up). It's hard for them to interpret abstract representations and, as a consequence, icons should be more literal than the ones designed for adults. In *Toddler Games* by Bimi Boo, each activity is represented by a tinier, slightly simplified, version of the main visual of the activity, and that is a good compromise between clarity and keeping the design not too busy.

Icon Design for Children

The first thing to notice about designing icons for kids is that, as mentioned in Chapter 3 and then again in Chapter 7, they develop fast. An icon designed for a 20-year-old user works perfectly fine for a 60-year-old. That's not the case with children. What you design for a 12-year-old might not work for a 5-year-old. And vice versa, because while the older kid surely understands an icon designed for a younger age, we need to take into account that age-appropriate styling is important (we'll see more on this topic at the end of the chapter).

In the beginning of this book, we saw how it's almost impossible to design an app able to appeal kids of all ages. Apps that target a wide audience, like YouTube Kids, should provide a different experience according to the age of the user. YouTube Kids partially does that, but the icons are the same, from preschoolers to 12-year-olds.

In Figure 8-25 we can see a graphical representation of the relationship between age and level of abstraction in an icon design. This, of course, can't be taken as mathematically exact; it serves just as a visual aid to understand the idea.

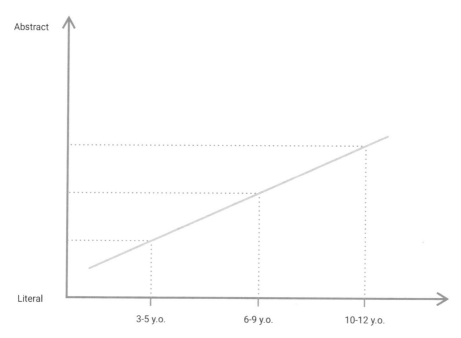

Figure 8-25. Relationship between age and level of abstraction allowed

Icons used in products for children that are 6 and up can also be simplified, thanks to the fact that we can start using text labels to go along with the icons. We shouldn't rely on text only, or too much, but we can afford to start being less literal and detailed in the designs.

One very important thing about icon design for children is the use of mental models they understand. Earlier in this section I made the example of the floppy disk icon and how young people probably won't understand where that is coming from. This is true (even more) for children that have a limited knowledge of the world compared to adults.

In the following exercise (Figure 8-26), I imagined how an icon to represent always the same functionality would evolve as the user grows up.

| 3-5 y.o. | 6-9 y.o. | 10-12 y.o. |

Figure 8-26. Evolution of an icon according to user's age

Let's analyze the icons one by one and see the differences in each step in Table 8-1.

Table 8-1. Icon comparison

In the icon for a 3–5-year-old:	In the icon for a 6–9-year-old:	In the icon for a 10–12-year-old:
• Bold colors and high contrast	• Fewer elements	• Even simpler design
• Dark outline on all elements to increase contrast	• The base does not have a strong outline	• No dark outlines, but the art still has a "childish" style
• Shadow behind the drawing tools, to separate them from the background	• No shadow behind the pencil	• Pill-shaped container, more similar to apps for adults
• Chubbier and "cute" proportions	• Thinner, less "cute" proportions	• The marker is straight, looking more "serious"
• No text	• Big text; use of a label with a short simple word	• Smaller text size; longer word for the label, with a more mature word choice; label positioned underneath, more similar to apps for adults
• Big squared touch target	• Small touch target	• Even smaller touch target

Even though the icon for younger children (on the left in Figure 8-26) involves more elements, it's important to notice that it still has a limited color palette and doesn't go overboard with details, and it's far from being realistic (skeuomorphic, remember?).

It's important that each element that makes the icon is easily readable and clear. Exaggerating with too many shades of color, too many details, trying to achieve a realistic look is not only useless but could even be detrimental.

Sometimes You Have to Follow the Standards

There are situations where trying to be literal can be very complicated, if not impossible, and even counterproductive. One classic example is the "play" icon, used both on media players to start a piece of content, and also used in games to start playing. Trying to come up with a literal representation for both these actions is not just a complicated matter, but also detrimental, in my opinion. The reason is that this kind of visual code is the standard, not just in digital interfaces for kids, but for any interface in general. Therefore, it's better for kids to learn what this symbol does, even if it'll be obscure in its meaning the first time they'll see it. Same goes with the "pause" icon and other similar standardized designs (Figure 8-27).

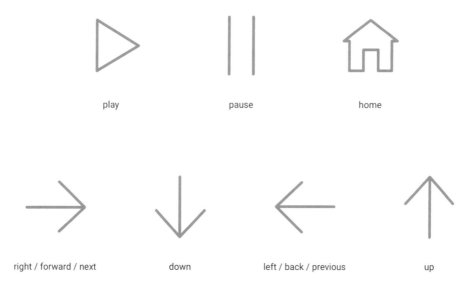

play · pause · home

right / forward / next · down · left / back / previous · up

Figure 8-27. Some examples of standard icons

The "home" icon is a particular case, because it represents a real thing to describe a nonphysical space. Even young children understand that drawing is a "house," but associating that to the idea of a home screen is an abstraction that can be impossible to grasp (older ones, as they get confident with web navigation and digital products, will understand the idea). Currently though, many children's products use it. My suggestion is to have an information architecture that is flat enough to require 1 or 2 taps on back buttons to go to the home screen (see Chapter 7), without the need to a shortcut using the house icon, in products for 3–5 years old. Older kids can deal with slightly more complex products and architectures and they'll understand soon enough what that icon means.

At the end of the day, you have to consider that children won't use exclusively your product. They'll use it along with many others, and reinventing the wheel defining a different visual code will just make your product odd and frustrating to use. Don't underestimate the capability of kids to learn and memorize new shapes and their meaning. Even if they won't understand at a glance what an icon means, they'll learn very quickly.

When designing icons for kids, we should keep in mind a few things:

- Rely on age-appropriate mental models.

- Use bold colors, but not too many.

- Contrast is very important.

- The younger the kids, the more literal the icon should be.

- Literal doesn't mean realistic. Too many details can be confusing.

- Think about the size this icon will be used at. Smaller touch targets, for older children, mean more minimal designs for the icons.

- Don't reinvent standard icons. Children won't use exclusively your product.

Typography

A very common misconception is that the typography on children's product has to involve the use of quirky, funny, practically unreadable typefaces. This is wrong for several reasons. First, ugly fonts are still ugly also on children's products, and that might be just a matter of personal taste and sensitivity about typography. There are other reasons though and taste has little to do with them.

If we're designing for preschool and elementary school children, we have to be careful because these are their first encounters with reading and writing. While it's very easy (well, depending on the typeface) for adults to recognize an "a" in different styles (Figure 8-28), seeing it in cursive and print, serif and sans serif, very simple or very ornamented might confuse kids.

Figure 8-28. Different styles to write an "a"

A font that is difficult to read or differs too much from what they are learning at school can be confusing and frustrating. It could ultimately even discourage kids from reading. That's not what we want.

This has to be considered even on products intended for nonreaders. Your app could be targeted to very young children who cannot read yet (and we talked about solutions on this topic earlier in this book), but it's still a good idea, even for the little text you might have, to be written in a font that trains their brain in recognizing letters, without struggling and creating visual noise.

There are different opinions on what makes a good typeface for children.[12,13] We base many of the most common assumptions on intuition, practical use, and tradition by educators, but there is no scientific approach to really validate them. Consequently, many choices rely on hypothesis and it's hard to really provide a set of rules that is backed up by data.

The choice of style for a typeface for children is often based on the idea that this should be consistent with graphical approach for handwriting taught in schools. But some educators challenge this approach, with the argument that the typeface should help kids familiarize with typography found in regular books and magazines. Both seem valid reasons.

Many type designers and type foundries came up with typefaces made for beginner readers. Most of them took the traditional rules as a starting point for their design, for example, it's widely assumed that *sans serif* fonts are easier to recognize by children. Sans serif fonts are designed "bare bones," the form of the letters is made only by their skeleton, it's the way we normally write when writing in *print*, we don't add serifs to our letters. The design of children-friendly typography went even further with the introduction of the so-called infant characters. These characters were introduced in the 1920s and became more popular during the 1930s. What does infant character mean? These characters are variations on specific letters made to make them easier to read for children. The most common examples of such letters are single-story[14] "a" and "g" (Figure 8-29). Here's the difference.

---

[12]Bessemans, Ann. "Typefaces for Children's Reading." *TMG Journal for Media History Typografie in Mediahistorisch Perspectief* 19.2 (2016): 1. Print.

[13]Wilkins, Arnold, Roanna Cleave, Nicola Grayson, and Louise Wilson. "Typography for Children May Be Inappropriately Designed." *Journal of Research in Reading* 32.4 (2009): 402-12. Print.

[14]Single-story letters have only one counter, while double-story ones have two. A counter is an enclosed or partially enclosed (usually) circular or curved negative space, for example, the "eye" of the lowercase "g".

single storied   double storied

Figure 8-29. Single- and double-story "a" and "g"

Other common examples of infant characters are the uppercase "I" and the lowercase "i" (Figure 8-30) that in many sans serif fonts look almost identical. In their infant version, the "I" has serif and the "i" has a curved terminal.

Figure 8-30. Regular sans serif "I" and "i" compared to infant "I" and "i"

Type foundries introduced these variations on popular fonts such as *Gill Sans* and *Bembo*. Type designers, like Rosemary Sassoon and Adrian Williams, designed typefaces specifically for children's book. Their typeface *Sassoon Primary* includes all the infant variations of the characters, including the curved terminals of "I" and "t" and the serifs on the capital "I". It's a very good typeface that I like a lot.

Serif vs. Sans Serif

The choice between *serif* and *sans serif* is also a matter of medium. The 2016 study[15] by Berrin Dogusoy, Filiz Cicek, and Kursat Cagiltay, *How Serif and Sans Serif Typefaces Influence Reading on Screen: An Eye Tracking Study*, is just one of the studies that highlighted, through testing with eye-tracking technology, how sans serif typefaces are more readable on screens compared to serif ones.

The 2008 study by Sheree Josephson *Keeping Your Readers' Eyes on the Screen: An Eye-Tracking Study Comparing Sans Serif and Serif Typefaces*, published in

---

[15]Dogusoy, Berrin, Filiz Cicek, and Kursat Cagiltay. "How Serif and Sans Serif Typefaces Influence Reading on Screen: An Eye Tracking Study." *Design, User Experience, and Usability: Novel User Experiences Lecture Notes in Computer Science* (2016): 578-86. Print.

Visual Communication Quarterly, got to the same results. This study is particularly interesting because the author didn't just compared serif vs. sans serif typefaces, but also compared fonts that were made for print with others made for screen. Overall, the best results were obtained with the use of a sans serif font specifically made for screen (in that case *Verdana*, designed by Matthew Carter for Microsoft in 1996).

In a 2012 article[16] on NNG, Jakob Nielsen challenged this guideline about the preference of sans serif fonts for online reading. His (valid) argument is that the "rule" was formulated when screens were low in resolution, hence the rendering of fonts with serifs was far from ideal, making texts less readable and blurry. In relatively recent times, screens got a lot better. Long gone are the days of Verdana and displays with 72 dpi resolution. So, if serif fonts are considered more readable on print, shouldn't they be considered the same on screens with a resolution comparable to printed pages? The 2016 study by Dogusoy, Cicek, and Cagiltay mentioned earlier seems to disagree, but that might not be enough. The debate on whether serif or sans serif typefaces should be the choice for online reading is still on today, but, bringing the conversation back to the topic of this book, there is one important thing to notice: digital products for children don't usually require long paragraphs. Quite the contrary, actually.

It's undeniable how serif fonts generally look more serious though and also how on digital interfaces sans serif fonts look more modern and "techy" (there is currently no operating system out there, desktop, mobile, or else, using a serif font as their default choice). Digital products are still very much tied to the "old" rule using sans serif font families.

In the end, sans serif font seems the best choice when it comes to digital products for kids, being them web sites or mobile apps.

What Are the Characteristics of a Good Font Family for Children?
Round Counters

Simple, friendly shapes, generously drawn with rounded *counters (Figure 8-31)*. Avoid counters that are rectangular or too narrow. Wider typefaces provide a better readability and they also resemble more the style of handwriting children learn at school.

---

[16]Nielsen, Jakob. "Serif vs. Sans-Serif Fonts for HD Screens." *Nielsen Norman Group*. N.p., 1 July 2012. Web. 23 May 2020. <https://www.nngroup.com/articles/serif-vs-sans-serif-fonts-hd-screens/>.

⊗ Apps for all kids

✓ Apps for all kids

Figure 8-31. Counter's roundness comparison

Taller X-Height

The x-height of a typeface is the height of its lowercase "x", which is the height that should guide the design of all lowercase letters without their ascenders or descenders (Figure 8-32).[17] Typefaces with a taller x-height look more readable.

⊗    ✓

Figure 8-32. X-height comparison

Avoid Decorative Typefaces

Decorative typefaces should be avoided at all times on any digital product, regardless of the age of the users. But it's very common to think overly decorated fonts are childish and therefore good for kids' books, websites, and apps. This is simply wrong for all the reasons mentioned previously. Such typefaces are not fun to read (or watch), just confusing and overwhelming (Figure 8-33).

---

[17]The ascender is the part in a lowercase letter that extends above its x-height, for example, the stem in a "b" or "d". The descender is the part that extends below the x-height, for example, the stem in a "p" or "q".

⊗ ‿Apps for all ‿Kids‿

⊘ Apps for all kids

Figure 8-33. Decorative typeface compared to simple sans serif

Pay Attention to the Font's Weight and Balance

The weight and width of a font have a direct impact on its readability. Fonts that are too bold or too thin, too condensed or too extended, tend to be less readable than a well-balanced font (Figure 8-34).

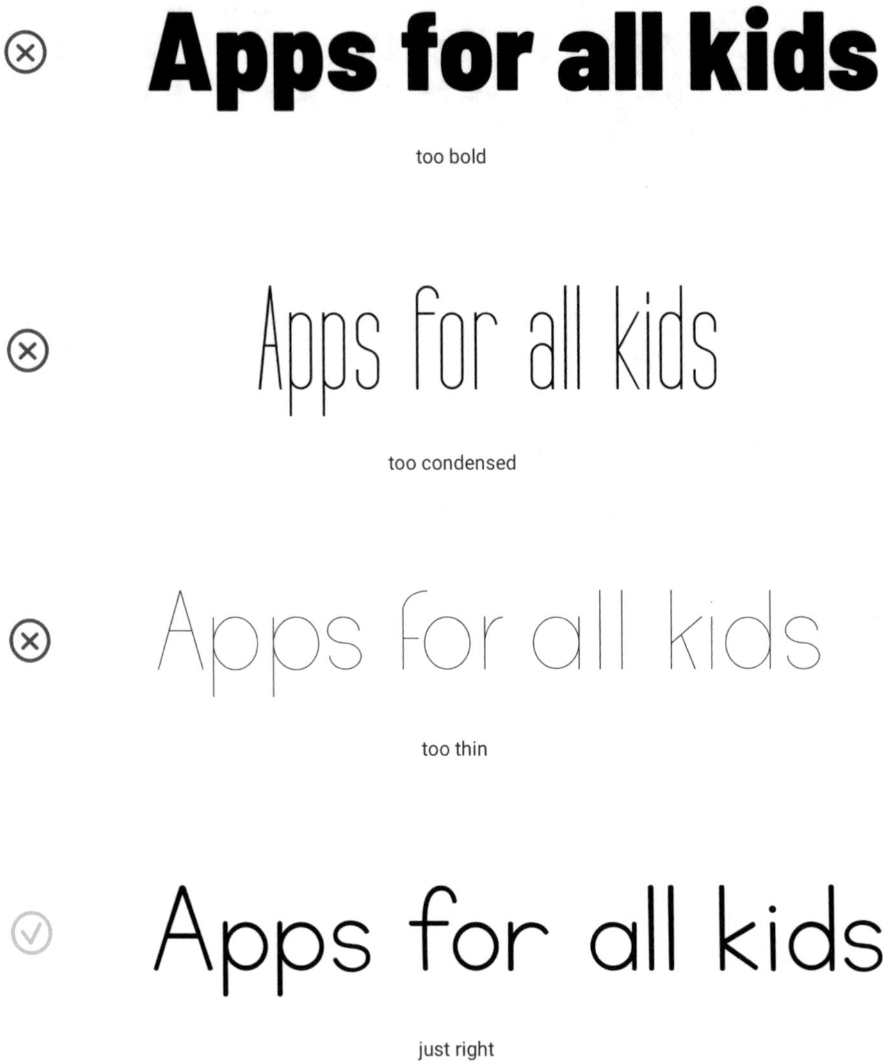

⊗ **Apps for all kids**

too bold

⊗ Apps for all kids

too condensed

⊗ Apps for all kids

too thin

⊘ Apps for all kids

just right

Figure 8-34. Different font weights in comparison

What Are Some Good Typefaces Then?

The choice of the right typeface can be time consuming. There is an uncountable number of options out there, and new ones become available on a daily basis. If you don't have much time to spend on this research, I can name a few options that tick all the boxes mentioned previously, but the choice should also reflect your brand and the look and feel you want your product to have.

Gill Sans Infant is a classic design, an old typeface, and one of the first to introduce and popularize the use of infant characters. *Sassoon Primary*, mentioned earlier in this section, is a more modern typeface and it has been designed specifically for children. It's important to notice, though, that neither of these typefaces was meant to be used on screens, but rather on printed material.

Kimberly Geswein is a type designer who made several font families for children. Her *KG Neatly Printed* is the font I used in the figures used in this chapter to show the correct version in each of the points. I like that because it's, as the name suggests, very neat, but friendly at the same time, not too serious.

Font Size

Size is undoubtedly one of the main factors influencing the legibility of a font. Early readers' material is usually characterized by large fonts, so that it's easier for children to distinguish letters.

The size of texts matters the most when we're dealing with digital products aimed to children in this "early readers" years, usually starting around 5 years old up until 11, while in products for younger children, not yet able to read, texts (if any, and we should try to keep them at minimum) are mostly intended for accompanying adults.

In early readers' material, the x-height of fonts usually goes gradually from about 4 mm/0.16 inches to roughly 2 mm/0.08 inches, which is adults' size of fonts. Pay attention here that I said "the x-height" and not the size of the font. The ratio between x-height and font size is different for each font; take a look at the following example to better understand (Figure 8-35).

Figure 8-35. Font size comparison on an iPad 10.2" (for proportion only; not to real scale)

As you can see, depending on the font we use, the size to get an x-height of 4 mm is very different, and this depends on how the font is designed. Now, an x-height of 4 mm is a very big text, but here we're talking about the proper sizing for paragraphs of text, for example, in an illustrated book. If we need to label some icons or add text to a button with a single word, we can go smaller than that.

Leading

Leading (in typographic terms or line height, if you speak CSS) is the distance between lines of text. As you can imagine, this is another variable we can play with to try getting better readability. Normally, font families should have a default leading that is good enough in most cases, but in regard to early readers we can increase the leading just a bit, to make the paragraph more airy.

There isn't an exact formula, also because it depends on the typeface, its ascenders and descenders heights, its x-height, and so on. But a rule of thumb I tried several times is taking the x-height of the font, multiply it by 1.618 (the golden ratio; you can do 1.5, it's OK, the golden ratio is mostly just a fetish) and consider that as the distance between the baseline of the line above and the x-height of the line below. That is usually (not always) a slightly bigger leading than the font's default value (Figure 8-36).

Figure 8-36. A slightly taller leading improves readability for early readers

Letter Case

Texts written in uppercase (all caps) are less readable than texts written in sentence case (starting with a capital letter followed by lowercase letters. Just like this sentence you're reading now) (Figure 8-37).

This is a general rule of typography, not specific to children, but considering we're dealing with early readers (if readers at all), it's one more reason to apply this best practice.

Figure 8-37. Uppercase sentences result less readable than sentence case ones

In the end, my suggestion is to keep it simple, with sans serif non-decorative typefaces, having a well-balanced design (not too narrow, not too wide, too short or too tall x-height, right weight, etc.). The specific choice on which font family should depend on the look and feel you want your product to have and your brand. Then just follow the usual rules of good typography for what concerns kerning, spacing, line height, and so on. There are some best practices when it comes to designing for children, but these never conflict with the general rules of good design, and this principle applies on everything, typography included.

Buttons

The most basic UI component we can think of is probably the button. Buttons can take many forms and styles, and they are one of the most versatile tools in digital product design.

A button can link to another screen or a modal, it can close it, can add or remove an item, start or stop a video or a music; it can basically trigger any kind of action in a product.

They also have the great advantage of requiring the most basic interactions: click or tap. You don't double tap on a button, you don't drag a button, you don't swipe or pinch. Just a click or a single tap and that's it. It works.

This makes buttons a very important component, since many of the interactions in your product will rely on them. The first question then is: what makes a button easy to use?

There are several factors, like its visibility, its position, its affordance, but the most important one is probably its size. The size of the button determines the *touch target* on touchscreens or *clickable area* on cursor-operated devices. These measures should take into account the motor skills of children at each physical development's phase. We saw, in Chapter 7, how the age of the child corresponds to a different level of dexterity, both in fine and gross motor skills. Younger kids require bigger targets, because of their less developed coordination. When they grow up, they refine and perfect their ability to control limbs and fingers, and our interfaces can rely on smaller targets.

Fitts' Law

In UX design, we have a principle called *Fitts' law*. Paul Fitts was an American psychologist who, in 1954, observed how the speed of a movement to move an object and the size of the object itself were related to the error rate in performing such action. In other words, the faster the movement and smaller the object, the higher the chance of making a mistake. This principle has been applied, years later, in interaction design, when interfaces became graphical and the pointing devices, such as the mouse, were introduced.

■ **Fitts' Law**   The time to acquire a target is a function of the distance to and size of the target.

Jef Raskin, in his 2000 book *The Humane Interface*,[18] talks extensively about Fitts' law and how this has been applied when designing the Apple Macintosh. One concrete example of how they applied it: traditionally, on a Mac, the main menu of an application is positioned on the top bar, at the edge of the screen. This makes it very easy to quickly move the cursor up and find yourself exactly over the menu, with no chance of going beyond that point (because there is no more screen to go). On Windows, the main menu of an application is contained within its window. This subtle difference makes it harder (or slower) to reach and click the menu on Windows than on Mac, because the user needs to stop the cursor precisely at the right height.

This example is not pointless. Consider this same idea in products for kids that involve the use of a mouse or trackpad. Try placing menus in parts of the screen where it's easy to stop with movements that don't require too much precision, for example, along the edges.

Fitts' law is absolutely relevant when we talk about buttons, both on touch and cursor-operated devices. The relationship between the size of the button and the ease of clicking or tapping it is even stronger when we talk about kids because of the considerations about development I mentioned in the beginning of this section.

Touch Targets for Children

Let's talk numbers. Alright, we talked about the theory, but what, in practical terms, is the best size for buttons in children's products?

Human Interface Guidelines for iOS indicate an area of 44 × 44 pt[19] as the minimum size for a touch area, while Android's indicate 48 × 48 dp. These are

---

[18]Raskin, Jef. *The Humane Interface: New Directions for Designing Interactive Systems.* N.p.: Addison-Wesley Professional, 2000. Print.

[19]"pt" (points) on iOS and "dp" (density-independent pixels or device-independent pixels) on Android are units used to solve the inconsistency of pixel measures caused by the many different pixel densities of today's screens. For Android 1 dp = 1 pixel on a 160 ppi screen, so on a 320 ppi screen, 1 dp = 2 pixels. On iOS 1 pt = 1 pixel on a 163 ppi screen. In this context, a 3 pixel difference is negligible; therefore, we will talk about dp regardless of the OS.

Sketch, Figma, and other UI design tools work with these units. The actual size of the assets is determined during the export.

the minimum sizes to consider for adults, but even for us grown-ups, a comfortable size is much bigger. A 2006 study[20] indicates an area slightly smaller than 1 cm (0.4") for comfortable and fast access. We need to point out a couple of things though: the study refers to one hand–held devices using thumbs (so a smartphone's scenario, rather than a tablet's), and it's an ideal size in an experimental setting; in reality, no app can afford that much space for each tappable area.

The perfect reference for this size, if you want to quickly have an idea, is the size of an app icon on an iPhone home screen. It's exactly 1 cm (0.4"). On an iPad they are a little bigger, roughly 1.2 cm (0.5").

Why talk about centimeters (or inches) for a digital product? The touchable area of a button (or any other interactive component) is unavoidably linked to its size in the real world, not just in the digital space. Our fingers are real and have a size we can't measure in pixels, the same is for the device, it's a physical, tangible thing we hold on our hands. Using centimeters or millimeters or inches is a quicker and clearer way to understand the dimensions we're talking about. Once determined their appearance in the real world, we can convert the size to dp and then export according to the pixel density we need.

Usability tests conducted by NNG determined a minimum tappable area, for young children, four times the one considered comfortable for adults (1 cm), so a 2 × 2 cm (0.8" × 0.8") area.

Converted to dp, these measures correspond to roughly 64 dp (it would be 63 dp actually, but it's always better to round to even numbers) and 126 dp, since 1 mm = 6.299 dp (Figure 8-38).

---

[20]Parhi, Pekka, Amy K. Karlson, and Benjamin B. Bederson. "Target Size Study for One-handed Thumb Use on Small Touchscreen Devices." *Proceedings of the 8th Conference on Human-computer Interaction with Mobile Devices and Services - MobileHCI '06* (2006): n. pag. Print.

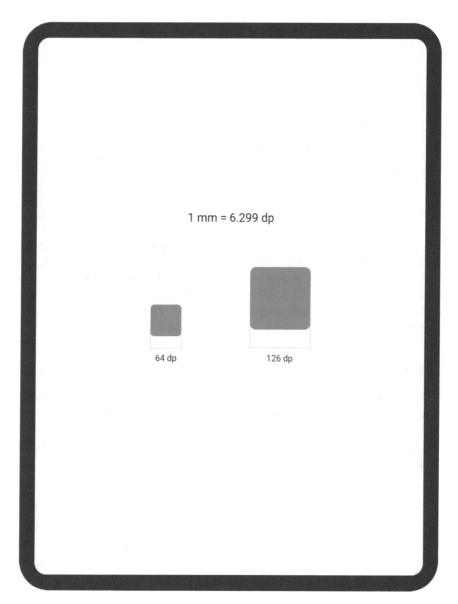

Figure 8-38. Comparison between adults' and kids' ideal touch areas

This big area is the ideal real estate we need for touch areas in products designed for preschoolers (3 to 5 years old). This size is inversely proportional to the child's age, the older the kid the smaller this area can be, up to about 12 years old, when we can finally consider adult's sizes (Figure 8-39).

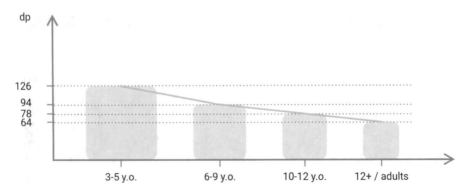

Figure 8-39. Ideal touch target size in relation with age

This relationship and measures indicated in Figure 8-39 are calculated by taking the ideal size for children as the starting value and the size of adults as the final one, and finding middle points for the intermediate age groups. But humans are not robot, so the motor skills' development is not as linear, we don't develop by downloading updates from the cloud (for now). Therefore, these should be considered as a rule of thumb and not mathematically.

Safe Space

When talking about touch targets, we should also consider the spacing between them. Two touch targets that are generous enough in size, but too close to each other, may be the cause of unintentional taps. As a rule of thumb, I usually consider an area 50% bigger of the touch target size as a safe space around it (Figure 8-40).

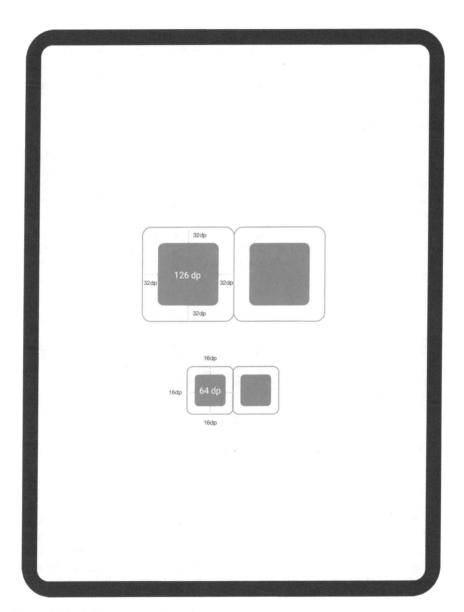

Figure 8-40. Safe space around touch targets

Buttons Styling

We already talked about skeuomorphism earlier in this chapter, and we saw how going there is not just unnecessary and out of fashion but, in some cases, even detrimental for the ease of use.

Styling is something strictly related to the brand, the product's personality, the look and feel you want the product to have. For this reason, it's impossible for me here to give any specific direction. Nonetheless, let's take a look at a couple of considerations. You'll decide if they make sense for your product or not.

Round Corners

In physical products for children, there are no sharp corners. The reason for that is, clearly, to avoid causing physical harm to the child manipulating such objects. Of course, in digital products, we don't risk such thing, but round corners are still a good idea (the roundness can vary, from a few dp to fully round in pill-shaped buttons; see Figure 8-41), because they look much friendlier than sharp ones (see Figure 8-41 for a comparison). Right now (in 2020), this is also a trend in products for adults, but for kids is a no-brainer choice, and I highly doubt we'll see a trend where sharp corners are hip in children's products.

Figure 8-41. Different degrees of corner roundness

Shadow or Not?

A middle ground between flat and skeuomorphic is the style used by Google's Material Design. In this design language, components are flat, but they work similarly to pieces of paper resting at different elevations and casting shadows on the elements underneath (Figure 8-42). The illusion created by this fake 3D space helps users to understand hierarchy and also to distinguish what is tappable or clickable.

With young children though, this does not work. They don't perceive things on screen as a 3D space, so shadow or not, it's flat for them. This doesn't mean you can't use shadow, but, at the end of the day, is more a stylistic choice than a UX one.

no shadow	soft shadow	solid shadow

Figure 8-42. Different kinds of shadows. For kids there is little to no difference

Avoid "Fake Buttons"

Buttons should look like buttons, that's a simple rule. But another rule is that non-buttons should not look like buttons. Watch out for elements in your UI that might resemble buttons, but are not interactive elements; for example, a container for a score counter should be clearly different from something you want to tap to get somewhere or to perform some kind of action (unless it does) (Figure 8-43).

fake button	simple counter

Figure 8-43. On the left: a score counter with a button look that suggests an affordance. On the right: a simplified version that does not invite the user to tap on it

Other Components

Usually, on children's products, we don't have the same wide library of components we can find in products for adults. Things like radio buttons, check boxes, scrollbars, drop-down menus aren't normally used in products for children, especially preschoolers. Most interactions are based on taps/clicks on buttons and occasionally drag-n-drops and swipes.

Keeping in mind the best practices for touch/click targets and all the others we just discussed in this section, as well as the information in Chapter 7, it should be rather straightforward to understand which components we can use and how to design them.

Fundamentals of Character Design

This might look like an unusual topic for a digital product design book, but in this particular context it's not. Characters play a huge role in apps for children, and I dare to say it's almost impossible to find one that doesn't feature any character at all. Their role can be different from app to app, but they're ubiquitous to any kind of product, from educational to entertainment, and everything in between.

There are apps for adults that use characters in their experience, but the majority (video games aside) don't. Characters are a peculiarity of digital products for children and they serve real purposes, they are not just aesthetic meaningless additions.

Why having characters in your product?

- To create a stronger emotional bond between the user and the product.

- To provide feedback, guiding the child through the experience.

- To communicate in a clearer and more efficient way.

- To make the experience more fun.

- To create a stronger brand and open it up to the possibility of bringing it outside of the digital world (or app series).

All of the above are great reasons to add one or more characters into your digital product. They serve both a UX need and a branding/business one. Characters in a digital product for children are an integral part of the UI.

From a marketing standpoint, characters can really help making a brand stronger and a product successful. Much of the success of *Angry Birds* can be attributed to a good character design, in my opinion. Sure, it's a fun game, but not even close to a new concept. Characters (and storytelling) contributed in making it that big hit and a multimillion worth brand that includes feature films, a theme park, and an endless catalogue of merchandise. The same idea without characters, or with badly designed characters, wouldn't have reached the same level of success.

Another great example of successful characters in digital products, and, in this case, specifically for kids, is the *Toca Life* series by Toca Boca. The characters from this series of apps (each one dedicated to a theme, like *Toca Life: Vacation*, *Toca Life: Farm*, *Toca Life: Hospital*, and many more) are used on an H&M clothing collection for kids.

Now, you might say: what do you know about this? Well, it happens that besides my career as a digital product designer (for children and adults), I've also worked as illustrator and character designer on the side, for quite a long time. I illustrated characters for many clients around the world, such as Nike, Foot Locker, Hasbro, and many more. In the educational field, I worked for ESL books by Oxford University Press and others, but, maybe, the most interesting project has been the one I did in 2017 for BBC's children TV network CBeebies. The project was for an educational series called *Kit & Pup* (Figure 8-44) created by the BBC's in-house production team. This last work is particularly relevant because it started as a TV show (lasted 52 short episodes of 5 minutes each), but later expanded into a digital product in the form of a *Kit & Pup* educational game.

Figure 8-44. Characters I designed for Kit & Pup TV show

In this section I want to cover the main steps to create a good character for your product.

Define Its Personality, Name, and Story

Even in the case you won't actually show your character story within the app, as it will serve more as a mascot, it's important to define the personality and a bit of background about who the character is, as well as giving it a name. The personality will affect the way the character looks, the way it moves, its expressions, how it talks, how it communicates with the user, and so on. Shape it around your brand and the concept of the app. What do you want to communicate? How do you plan to use the character? Is it a mentor to the user? Or a companion? Is it more an older sibling kind of figure? Is it a hero or someone the user should help?

Characters are based on archetypes and stereotypes. The modern archetypes, used in psychology and frequently used in creative writing, are based on the ideas of Swiss psychiatrist Carl Jung, who based his theories on Plato's work. It's not necessary to go too much into details on this, so please allow me to overly simplify in this way:

- The archetype answers the question "What does this character do?"
- The stereotype answers the question "How does she/he do it?"

Decide Its Species

Characters can be human being, animals of any kind, mythological creatures, aliens, artificial beings… Anything can be or become a character, even plants or objects!

Once you decided the personality and story of your character, I suggest to try several options, in the form of quick thumbnail sketches. The first task I was assigned when working on the project for BBC has been an exploration of a character duo, knowing only that the show would be all about the world of opposites (hot/cold, wet/dry, etc.). I sketched many options, from animals to humans to geometric shapes to fantastic creatures, for a total of 64 different characters (some of them in Figure 8-45). During this phase, I didn't draw characters standing there, doing nothing, but I imagined little dialogues between them, situations where they are dancing, fighting, running, screaming, singing, and so on. This helped me to reinforce the idea I had for their personalities.

Figure 8-45. Just a few of the many thumbnail sketches I made for Kit & Pup

Establish the Proportions

Proportions depend on the feeling you want your character to provoke and communicate a lot about the character's personality. The ratio between head and body is one important variable to define the cuteness of the character. Characters with big head and small body look more cute, because they resemble the proportions we see in babies; this is the case of my characters for *Kit & Pup*, for example.

While if we want to make a superhero kind of character, we need more realistic proportions (Figure 8-46).

Proportions will need to be consistent, with the exception of occasional squashes and stretches in exaggerated poses to convey particularly strong emotions.

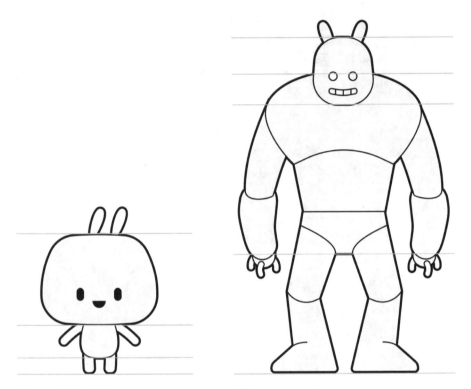

Figure 8-46. Different proportions convey a different personality

Then the character's *turnaround*, another basic step in character's development. A turnaround is a sheet showing the character from at least three different angles (e.g., front view, 3/4, and side). This gives an idea of the volumes and serves as a reference for all the poses (Figure 8-47).

Figure 8-47. Character turnaround from three angles

One thing you want is the character to be "readable," meaning that its shapes and volumes are easy to understand for the viewer. A way to check that is by looking at its silhouette. You should be able to get an idea of what the character is just by looking at it (Figure 8-48).

Figure 8-48. Silhouette of a character

Colors, Details, and Accessories

The amount of colors and details depends on the style of your choice, but in general, especially for characters in products for preschoolers, I suggest to keep them quite simple.

For the color palette choice, we can refer to what we discussed in the "Color Palette" section of this chapter. Bright colors, high contrast, limited palette are a safe choice in most cases.

Details and accessories are where you can highlight the personality of your character and make it memorable and distinguishable (Figures 8-49 and 8-50). Also in this case, it's better to keep in mind the age of our users and use accessories that they know and can identify.

Stereotypes can help. While applying stereotypes to people is a bad thing, on characters they help to send a message and communicate the personality of the character. When I had the idea for the character of Eli, in *Eli Explorer* by Colto, I thought to give him glasses, for two reasons: first, to show wearing glasses is OK and normal, even cool (that's what Harry Potter did for millions of kids); second, to give the idea of Eli as a curious guy, looking for new things to discover. Is this a stereotype? Well, yes. When considering stereotypes, we have to be careful and avoid perpetrating harmful and wrong ones. An innocent stereotype is nerds wear glasses (note that this is different from "people wearing glasses are nerds"); a wrong one is dancing is for girls. Other examples of innocent stereotypes: woodsmen wear plaid shirts and have beards, wizards wear pointy hats, owls are wise, and so on.

Always avoid gender and racial stereotypes at all costs! If anything, characters should teach about inclusivity.

Figure 8-49. Character finalized with colors and details

Figure 8-50. Different tests for the ears of a character for Kit & Pup

Dynamic Poses

To really understand if a character works, dynamic poses are essential. Seeing a character in action is important to define how its personality is reflected in the way it moves and also to check its proportions. Dynamic poses should also include the use of items, to understand how the character grabs objects or rides vehicles, for example (Figure 8-51).

Figure 8-51. Dynamic poses test

Exaggerated Expressions

The main reason to have characters in your digital product for children is to create an emotional bond and communicate more clearly with the user. For these purposes, one thing to keep in mind is that your characters have to be expressive. Their feelings and reactions have to show clearly on their faces and overall body. Exaggerating the size and deformation of facial features is one way of doing this. It's also important to make the expression very different from one another; subtle differences might go unnoticed. Laughs need big open mouths, surprise big wide open eyes, and so on; also consider the use of classic props such as heart-shaped eyes and similar (Figure 8-52). If animated, the motion should also emphasize the change in expression.

Figure 8-52. Different expressions of a character. Facial features are exaggerated to emphasize the emotion

Simplicity

A character used within a UI should be simple enough to not distract users from the tasks they want to perform (or you want them to perform). Its role is to guide the child through the experience, give feedback, establish a connection, and suggest an emotional response. It's a supporting role, not a main one. The child should be the main actor in the experience, the character is there to help and entertain.

Especially when you design for preschoolers, another good reason to keep your character simple is to make them easily understandable by the children and possible to be drawn by them. Take, for example, *Peppa Pig*. Peppa and all the other characters in the show are made of simple shapes and flat colors. The limbs are sticks, the body is a simple geometric shape: half ellipses for the females and circles for the males. Children can draw Peppa and her family and friends quite easily. I think this is a very smart feature for a character for young children.

Inclusivity

Characters can play a great role in promoting inclusivity. It's important then, as I was mentioning earlier, to avoid gender and racial stereotypes. This doesn't mean all characters should be gender neutral, but roles shouldn't fall into stereotypes straight from another era. Diversity should be celebrated and embraced.

Toca Boca pays great attention to this and the characters in their products are very diverse and never stereotyped in a bad way. The overall style of Toca Boca's apps is gender neutral; none of their products is clearly targeting males or females specifically.

But talking about inclusivity can't stop at gender and ethnicity. Inclusivity has to take into account also people with disabilities. A great example is a game called *Cardpocalypse*, available on several platforms, including Apple Arcade, Steam, PlayStation, Xbox, and Nintendo Switch. The main character in the story is a girl in a wheelchair, something that, unfortunately, is still very unusual in character design.

Age-Appropriate Styling

It's impossible to make a digital product aimed to kids from preschool to 6th or 7th graders, or, at least, really, really hard. As I mentioned many times in this book, children develop quickly and what is good for a 5-year-old is not for a child just a couple of years older.

Children want to feel empowered and treated in a way that is fair for their age. An app made for preschoolers can't appeal kids in elementary school because, with its huge buttons and super simple interactions, they would see it as a product for toddlers. It's the same thing as for physical toys. An 8-year-old won't play with one of those plastic toy phones with big chunky colorful buttons doing noises made for toddlers. All about its design says it's made for babies.

With digital products, it's not different; the experience you design on an interaction, but even more on a visual design standpoint, must speak the language of the age you're designing for. This influences all the points we've seen in this chapter, including colors, typography, design of components, and character design. Don't make big buttons thinking "if this is good for preschoolers, it'll be OK for older kids as well." From a usability standpoint for sure, it works, but from a marketing standpoint, it will be perceived as too babyish from older children.

Toca Dance and *Toca Hair Salon* are designed for an older audience than *Toca Pet Doctor*, for example, and the style of these apps is appropriate, in all their components, for these different age groups. Not only interactions are simpler in *Toca Pet Doctor* but also the use of colors, the style of the characters, the design of the interface. Attention to things like these makes developers like Toca Boca stand apart from the competitors.

It's important, then, to understand your target well and see what is appropriate for their age, what they like, what they use. To understand this you can look at competitors, sure, but try also to have a wider view, for example, by understanding what TV shows they like, what games they play, what are the latest trends in fashion even. Defining the style of your product is not just a matter of putting pixels together; it can make the difference between your app and the competitors, it can decide its success.

Industry Insight: Interview with Chris Bishop

Chris Bishop has been creative director for PBS and PBS Kids. During this time he worked on countless digital products for children involving many popular IPs. He started his career when products for kids were just at the beginning of their evolution and witnessed the changes in the industry for two decades.

Rubens: What has your first experience with a digital product for kids been? What kind of product was it? How did you approach it, without having prior knowledge on the subject?

Chris: In 2000, I was the first designer hired to work for PBS KIDS Digital. Our main focus was pbskids.org, the online home of games for all our TV shows. I was hired because the combination of illustrator and designer is a perfect fit for kids' design. My approach was to focus on fun, bright visuals. I didn't want it to look like a website with a bunch of columns and boxes. I wanted it to feel like a game. I didn't want it to look like anything else out there. Even now, the interfaces I create are a blissful merger of illustration and design.

R: Since then, how did the landscape of digital products for children change? Besides technological improvement, do you think the design side of the thing has become better and more conscious?

C: I think the ideas and creativity were always there. The technology has caught up so now we can actually do just about anything we can think up.

R: What are the three most important things to keep in mind when designing for kids? And what is the biggest mistake one can make?

C: 1. Think like a kid. This is actually very hard to do. As adults, we have preconceived notions about how things work. How many video player experiences have you used in your life? I'm sure you can easily list out their standard features and picture a typical layout in your mind. Now imagine this is the first time you are ever using one. How might you expect it to work? What features do you not need? How can using it be fun?

2. You have no idea how young or old your users will be. Older kids play games made for little kids and little kids play games made for older kids. Plan for that.

3. Respect the kids you are designing for. They are more sophisticated and clever than you might think.

The biggest mistake is getting too caught up in an idea that YOU think is cool that you lose sight of whether a kid will understand it or not. I am guilty of this all the time. It's easy to get overly excited about games and experiences that you as an adult like.

R: If you think about how kids were when you started your career, compared to kids 20 years later, do you think they changed in any way? If so, how?

C: For the most part, kids are the same. They like fun. They want to play games. On the other hand, the devices kids grow up with have gotten more and more kid-friendly over the years. We've gone from desktop computer with a mouse to portable kid-sized touchscreens and who knows what will be next.

R: In your opinion, what's the best way to conduct tests with children? How is it different depending on their age? How is it different from testing with adults?

C: Any type of testing is great. We do formal tests in labs and we do informal tests in school libraries. Getting real kids hands on with your product is so important, even if it's just your friends' kids from the neighborhood. I've never met anyone who knew 100% what was going to work with kids. It's impossible. They always surprise you.

Kids don't do that thing adults do where they tell you what you want to hear or try to look smart. I find kids have no trouble telling you how they actually feel about what you are testing—positive or negative. Also, it can be pretty clear what makes them perk up in their seat or what makes them slump and eyes glaze over. Kids don't feel the need to hide that for the sake of your feelings.

R: What's the most common flaw in digital products for kids today? How can we address it?

C: Rainbows. Everything doesn't have to be red, orange, yellow, green, blue, violet. Kids can handle a beautiful, minimal color scheme.

R: How important is it to have characters or a mascot in a digital product for children? I'm talking about custom-made characters as well, not necessarily from popular IPs.

C: Characters are everything. To kids they are real.

R: Last question: what's the secret to making a good interactive experience starting from a classic TV show? In other words, how do you make something that is traditionally speaking one-way, and enjoyed in a more passive way, interactive?

C: It's actually pretty easy. For many years, most kids shows have had built in interactivity like turning to the camera, asking a question, and allowing time for the kids at home to shout out the answer. Kids don't have a problem making the leap from TV to a digital platform with a character they love. Digital experiences allow them to continue to play with their TV pals but be in control—dressing them up, deciding where they go or what they do. Digital can open up the whole world of the show to the kid to explore.

Industry Insight: Interview with Chris O'Hara

Chris O'Hara is an Annie-nominated animator and character designer with over 10 years of experience. During his career he has worked for clients such as Cartoon Network, Disney, Google, and Netflix.

Rubens: One thing that goes often overlooked when talking about the design of digital products for kids is the importance of characters, when, in reality, most of the products for kids out there use them. Why do you think characters are so ubiquitous in apps for kids?

Chris: I would say the characters act as a guide, like a teacher or a parent would. A figure that can lead the children through the learning experience much the same way they learn from parental figures in real life. The characters can also be perceived as a friend for the child, to support them and encourage their learning in a fun and friendly way.

R: When designing a character, I think the age of the viewers is an important variable to take into account. What are the differences between designing a character for preschoolers and for older kids?

C: The main difference would be simplicity. Designs for preschoolers usually focus on simpler design and color. The designs often feature less details that could be distracting for younger viewers. Cute designs are usually more prominent in preschool characters too. Big eyes, cute proportions, traits that reflect the kids that will be watching and make the characters more relatable to that younger audience.

R: What makes a character design good?

C: To me its appeal, character, and clarity. The goal is to capture the character in its design. You should be trying to present personality in the design. If you get a sense for who this character is through the visuals alone, I think it's a successful design. There are certain formulated approaches to this including shape language, color choices, and exaggeration.

R: In a game called Cardpocalypse, the main character is a girl in a wheelchair. Do you think characters can help to promote inclusivity to children?

C: Absolutely! The goal should be to present all kinds of children in the content we produce. The world is full of diversity with a place for everybody, and the sooner children learn this the better. Featuring characters of all backgrounds helps to promote equality and nurtures children's empathy and respect for others that look different to them.

R: What's your process when designing a character?

C: It varies and there are many factors to consider depending on the project, but usually I try to learn as much about the character as possible. The more information I have the better I can try to produce an honest and appropriate design. I usually churn through as many rough ideas as I can and slowly it gets narrowed down and compiles the various ideas into one design that I feel best represents the character in an appealing way.

R: Characters can help to provide feedback, especially in apps for nonreaders. What do you advise to enhance the expressiveness of a character?

C: Clarity is important and clarity can come through exaggeration. Exaggerated design and expressions allow for clearly communicated thoughts and ideas. Elements like eyebrows offer a great range of emotion on characters and are an important design asset to communicating clear expressions and emotions. Clear and strong character posing is another way to communicate.

Chapter Recap

- The UI should help to distinguish what is interactive from what is not.

- Skeuomorphic design doesn't necessarily mean better affordances. In some cases it can be too much and make the experience less intuitive.

- Colors should be bold and bright, with the right amount of contrast.

- Saturated colors should be used for interactive components, muted colors for backgrounds.

- A color palette can be defined using different color harmonies. Each one can provide a different level of contrast between the hues in our palette.

- Products for kids require a wider array of colors than products for adults. Using too many though is not advised.

- In digital products for children, colors can help navigation.

- Consistency in colors is important. If you use a color to identify an area of your product, don't use it again for another area in a different screen.

- Avoid gender bias when using colors. Go beyond "pink for girls and blue for boys."

- When designing icons, try to be as literal as possible. Use mental models familiar to kids of the age you're designing for.

- Be more literal for younger kids. You can slowly move to more abstract designs as kids grow up.

- Standard icons (*play, pause, back, forward...*) should not be changed.

- Use simple sans serif typefaces and avoid decorative fonts.
- Use big fonts for early readers.
- Follow the rules of good typography used for adults.
- Touch areas for younger kids should be roughly four times the size of the touch areas for adults.
- The older the kids the smaller the touch areas can be.
- Add safe space around each target, to avoid unintentional taps or clicks.
- Rounder buttons feel more friendly.
- Don't make noninteractive elements look like they can be tapped or clicked.
- Consistency is very important. All instances of a component should always look the same (e.g., a *back* button can't do the same thing, but look different in different screens).
- Characters create a stronger emotional bond between the user and the product.
- Characters provide feedback, guiding the child through the experience.
- Use characters to communicate in a clearer and more efficient way.
- Characters make the experience more fun.
- Characters help to create a stronger brand.
- Use characters to add diversity and promote messages of inclusion in your product.
- The style of the product should be aligned, in all its components, with the age of its audience.

User Testing with Kids

Learn from Your Users and Perfect Your Product.

Testing leads to failure, and failure leads to understanding.

—Burt Rutan, American aerospace engineer

User testing doesn't necessarily come at the end of the development process. You can test an idea at the very beginning with lo-fi prototypes, even paper prototypes. And you should test along the way any time you think it would be beneficial, for example, to test a specific functionality or interaction before moving forward with it. For sure though, it's something you want to do before releasing to the market.

Also in this case, children are different from adults, and testing effectively with them requires to work according to their needs in order to make the best out of the testing sessions and get useful insights. It's easy to understand that doing user research with a 26-year-old and a 36-year-old is exactly the same in the way you conduct the tests. They might be in different age groups on a marketing standpoint, but usability-wise, they are two adults with a similar level of cognitive and motor skills. The thing is completely different with two kids with the same age gap. Conducting a research with a 3-year-old is a

© Rubens Cantuni 2020
R. Cantuni, *Designing Digital Products for Kids*,
https://doi.org/10.1007/978-1-4842-6287-0_9

completely different business than doing so with a 13-year-old. You'll be probably bored to death by reading this again, but let me reiterate one more time that even a 2-year difference means the world in children's development.

There are several things to take into account when doing user research with children, and in this chapter we'll try to cover the best practices for conducting user testing with them.

Recruiting

The age of the participants is a pretty obvious variable to consider; it needs to match the target age your product is made for. And, unless your product is specifically targeting one gender, you should also have a good mix of males and females.

When recruiting for user testing with children, it's a good idea to have a few more participants than you would normally have when testing with adults. There are a couple of reasons for this:

- When conducting tests with children, you have to deal with two people for each participant: the children and a parent/caregiver. This doubles the chances that someone might have a problem and doesn't show up.

- Sometimes children, especially the younger ones, can be shy having to talk with strangers, being asked questions, and so on. You can try all the tricks in the book (and we'll see some of them here), but sometimes there is nothing that works and you have to just let them go.

Besides the number of participants, variety in personality is also crucial. Of course you can't know them personally, but asking the parents if their child is more of an introvert kind of child, or more open, less or more talkative, or if maybe they tend to become more chatty when being together with a friend, and so on, can be useful to understand how to approach the kid and also to get a wider sample of reactions and responses to the product.

Don't forget the consent from the parents. Having to deal with minors, it's important to have all the necessary paperwork in order. Depending on the country (and the company) where you're conducting the test, the requirements might change, so be sure to comply with your local laws and regulations.

Environments

A welcoming environment is the first step to put participants at ease. Depending on your needs and possibilities, you can either

- Set up a room in your company or testing facility specifically dedicated to user testing with children.

- Find a public space, like a school or library, to use as testing lab.

The main difference here is that with a dedicated testing place you can have a better infrastructure, with different camera angles, two-way mirrors, and so on. The public space has the advantage of usually being a cheaper option, and if your participants are regularly going to that place (maybe it's their school or local public library), it offers the big advantage of being a familiar place, where feeling comfortable will be much easier for them.

In any case, this place must be designed with a children-centered approach. So the decorations have to be children-friendly, with colorful posters and other adornments. It's important to keep a good balance, between being warm and friendly to kids, but not too distracting. Try to avoid toys or other sources of potential distractions, unless the aim of the test is about playing with toys, of course. A window is good, but if the view offers too many stimuli, it's better to have participants not facing it.

Furniture has to be child centered as well, so chairs and tables should be of appropriate size for the age you're testing. Overall it's a good idea to make it feel a bit like a classroom, to instill the idea of proper behaviors.

Don't forget that you're part of the environment as well! So your appearance is also essential to make participants feel comfortable. Don't overdress, avoid formal attires (I shouldn't even need to mention lab coats. Please don't!), casual but professional outfits are the best choice. Another idea is to wear T-shirts or pins or stickers with popular characters among the kids of the age of your participants; it's a great icebreaker.

The whole point is to avoid making the children feel like they are being tested or feel like they're going to the doctor. Make it clear at the very beginning that the kids are there to teach *you* something, and not the other way around.

Parents' Presence

Getting parents' trust is important because children are more likely to participate and feel comfortable if they understand their parent or guardian trusts you. Sometimes, especially with younger children (2 to 4 years old), having the parent in the room can help to make them feel safe, but for other kids, having a parent present during the test could be detrimental, because of fear of being judged for giving a wrong answer (of course there are no wrong answers, but kids might perceive this as a possibility—try to avoid this at all costs). It's impossible to know what's better before actually trying, and all kids are different, so start with the parent out of the room, if the child allows it, and see how it goes. Read the body language; a kid that constantly looks at the door where the parent left is a clear sign that having her/him in the room could be a good choice.

If the parent or guardian is in the room, let the kid see where they'll be sitting, but then have the child face in another direction, to avoid her/him constantly searching for the parents' approval and to avoid the adult to influence the child with nods or other voluntary or involuntary messages.

It's essential to instruct the adult about being a silent observer. If the kid asks for the parent's help, you and the parent should explain that the child is the expert, the one teaching, and you just want to learn to encourage the child to try on her/his own.

Parents can also support in explaining the task or interpret the answers. Besides being a useful asset, it's a way to make them feel involved and gain their trust. But again, it's important they don't sway the responses in any way.

Friendship Pairs

Something I've noticed that works pretty well, especially with 6–8-year-old children, is the involvement of a friend (or sibling of a similar age). Working in a pair with a friend helps shy children to open up and speak their mind. It helps relieving the stress of having to deal with an adult stranger alone, and this sort of mini-brainstorming is an occasion to exchange opinions and hear from more than one kid at a time. The way they interact together with the product is also interesting, because it could spark ideas on new functionalities for teamwork and group play. What's important in this case is to avoid conflicts over who should use the device. Set turns and keep both kids involved so that no one is getting bored by watching the other playing.

The Prototype

Children are, by nature, more explorative than adults, so they're more likely to tap or click on anything just to see what happens. While a paper prototype could be good to explore specific playing scenarios at the very early stage of the concept, I'm a big believer in high-fidelity prototypes when it's time to really put your ideas to test.

The fidelity of the prototype should be as close to reality as possible, especially on the visual design. Adults can work on wireframe prototypes, if testing the visual design is not really our focus. A gray sharped-corner button that says "Sign up" is not so different from a more polished and beautified version of the same component, for a grown-up, if testing a sign-up flow is what we want to do. For a kid it could make all the difference. So prototyping interactions with their visual design is the only way to really get the right insights from your user testing with children.

The more flows, interactions, and screens you can include in the prototype the better. This would avoid the risk of the session getting stuck because of getting on a broken path.

A prototype though, even if hi-fi, is usually far from a finished product and not all flows and functionalities might have been designed and prototyped, and, while adults tend to stay more on track with the task, kids are more likely to tap on random things that might "break" the prototype. It's a good idea, then, to have a reset hidden somewhere (it could be a gesture, hard for kids to perform accidentally), to quickly go back to the initial state.

Asking the Questions

The first thing you want to make clear (also to the parents or guardians) is that it is not the child being tested, but the product. Let the kids know that they are there to help you and teach you how things work. Empower them by making them feel they are the expert, you are there to learn.

Before starting the session, it is best to have some icebreakers. Earlier in this chapter, I suggested wearing something with popular characters on, to spark a conversation about something they like (or dislike). That's one thing; you can also ask some general questions about what are their favorite games, or apps, what they like about them, if they use them alone or with friends or siblings, and so on. These open questions are easy for them to answer to and, at the same time, might offer some good insights.

You need to stick to very concrete and pragmatic questions. Don't be too abstract, as we discussed in Chapter 7; for children up to 11 years old (this is of course slightly different from child to child), it is still hard to think that way.

There are a couple of other things you need to clarify before starting the tasks:

- It's OK if they don't know how to do something or how to answer a question. They can seek for your support at any time.

- They are free to express themselves without fear of hurting anyone's feelings. If they don't like something, it's OK to say so.

The progression of the tasks' difficulty should be upward—starting with the easier tasks to help the children build confidence, then moving on to more complex ones at the end of the session. It's OK if the first couple of tasks are very basic and maybe not even part of your focus for that test. What's important is to ease in the participant. All tasks should be realistic in their complexity, but that should come with the concept. If you can't test a functionality because it feels too complex, it shouldn't be part of your product to begin with.

The language you use should be appropriate to the age of the children you're testing with. By "appropriate" I don't mean without swear words, that should be implied. I mean easy enough to understand for children of the age of your testers, while not being patronizing at the same time. This should be done both when providing written instructions, for older kids, and when giving instructions verbally, for nonreaders.

Children can easily be distracted and take another direction from the path you planned for your test. While this could be time consuming, compared to testing with adults, it's also important to let them explore (as much as the prototype allows that). You might discover things you didn't expect.

Duration

It's wise to always account for more time when testing with children, compared to testing with adults. As mentioned previously, kids are easier to be distracted and take their own way while doing a task. While this is not necessarily a bad thing, be sure to allocate a longer testing time.

The duration of each session should not be over one hour for each participant, to avoid losing their interest. Consider also having one or two breaks during the session, and I advise proposing a bathroom break right before actually starting the test (after some ice-breaking talk); this could prevent a source of distraction and unplanned interruptions later.

It's always a good idea to explain how the session will work, breaking it down to the activities you're going to do together, what device you'll use, what kind of questions and tasks they'll be asked, and so on. Having an idea of what to expect is another way to put your participant at ease and avoid them to feel anxious.

Feedback

Here's another key difference of user research with children and with grown-ups. When working with adults, you want to be as neutral as possible in your feedback and reactions. There is no right or wrong answer; the participants are testing the product, not being tested, so they are the ones telling what they feel is right or wrong with the product. There is no stick and no carrot.

While this last statement is still true for user tests with kids, your participants might need to smell a carrot to get motivation and gain confidence. Positive feedback is essential but has to be given carefully. You don't want to sway the participant to any direction in particular, so try to find ways to give a positive response without influencing her/his decisions. Try things like "Thanks for teaching me that!", "That's really interesting!", or "You're being really helpful!". Reward the participants' behavior, not their decisions. Positive feedback like this is a way to make children more confident and willing to speak.

Getting the Answers

Sometimes, it can be hard to understand what kids mean. Younger ones still have a limited vocabulary and their ability to articulate ideas is also lacking. There are a few strategies you can try to put in place to alleviate this problem. First of all, video recording the sessions can be helpful to review them later to listen to the answers again (multiple times, if necessary), but also to notice more subtle reactions in the participants' body language. If you do so, remember to clearly declare it in the consent the parents will have to sign, including references to all the necessary privacy policies.

It also helps if you can have someone else present to take notes while you focus on the kid. Giving all your attention to the child, without interruptions to annotate the answers, will benefit both the relationship you'll establish with the participant and the quality of the notes, given that someone else will take care of it.

One great strategy to facilitate answers from younger children is using a visual aid to express feelings. This is very similar to those posters we often find at a doctor's office, with different face expressions associated to pain levels (hopefully your product won't be causing those same agonizing reactions). It can be a piece of paper on which the kid can point at the expression that better represents their feeling (Figure 9-1), or they could be stickers that the child can place somewhere, to make it more fun (if you already have a character designed for your product or brand, you can use it for this purpose; see Chapter 8).

negative ⟶ positive

Figure 9-1. You can use emoji-like stickers to help the children express their emotions

Lastly, if a parent is present (see the "Parents' Presence" section earlier in this chapter), you can ask to chime in to help facilitate the answer, minding that the parent shouldn't suggest anything, just encourage and interpret the answer, if needed.

Industry Insight: Interview with Martina Dell'Acqua

Martina Dell'Acqua is a game designer focusing on products for children. She's Head of Product at Colto, a company specialized in educational games for children, partnering with big media companies such as Nickelodeon and Highlights.

Rubens: How do you handle recruiting? Is there anything else, besides the age, that you look for when recruiting participants to test one of your products?

Martina: Age is a key requirement for us but it is not the only one. I remember when we started 5 years ago; back then, it was extremely important knowing if the kids we were testing were using a mobile device. There were a lot of kids that were not used to these devices and it was fundamental for us to test both groups. Nowadays, it is extremely rare to find children that have never used a device. However, we still consider it important to test our apps with kids that don't have a strong previous digital experience to understand if our product is intuitive enough.

R: In your opinion, what's the best way to approach kids when they come for a user test? What's your secret to put them at ease and make them feel comfortable and safe? Can you share any tips and tricks?

M: Opening and giving the app straight away to the kids give useless results.

They freeze with their wide eyes open and stare at you. "What should I do?" they seem to ask.

My suggestion is to try to "break" this image of you as "the adult" and become "the friend." What I usually do is to just take my own device and start playing different games to spark their interest. Then, I start talking with them about favorite apps and TV shows to put them at ease. After a while, I ask them to try this new "app", the one I have wanted them to try since the beginning!

R: How important is the testing environment? What's the best setup in your opinion?

M: The environment is probably the most significant thing during a test with kids. Not the environment as "the building" per se, but the entire experience that you are having your test in. It is important to recreate the comfortable atmosphere in which they will use your app. In my experience, we tested in both schools and during playdate in our office. It is obvious to say the latter is a much easier way of testing it, kids are more relaxed and confident during a playdate that feels more like home. In order to achieve such atmosphere, in addition to digital products we like to have some physical games products that kids are familiar with, so they can use them and play together to build up their confidence and improve the overall experience.

R: And what about parents? How do you make them feel comfortable about their kid being involved in user testing?

M: I used not to interact much with parents. Recently, however, we begin to involve them much more in the user testing itself. We need to understand how much they are involved in their children's play and how. Of course, kids are the main users but the parents are important actors as well. They are the ones who buy the games, but most importantly they are often the ones interacting and playing with the kids. Making them part of the test results in making them much more comfortable with seeing their kid playing with the app.

R: Sometimes children, especially the younger ones, can have a hard time expressing their ideas or feelings about a product. What's your advice to help them in this situation?

M: It is very complex to understand what kids like and want. We usually rely more on nonverbal communication. Direct questions rarely work as kids tend to try to give "the right answer." As a workaround, during the years we came up with different questions that are easier for kids to understand and provide a more realistic answer. For example, "Do you like this game?" becomes "Do you want me to open another app?". "Describe this game" becomes "What would you tell your friend about this game?".

R: When testing with adults, we normally avoid giving any kind of feedback, positive or negative. While it's true that there are no right or wrong answers during a test, I believe it's also important to give some positive feedback to the participant, when working with kids. How can we give positive feedback and build their confidence without influencing the outcome of the test?

M: As you say, there is no right or wrong during a test, so the goal of the researcher should be "to break the fear of making a mistake." And this is true for adults but also it is especially true for children. In this case, what we like to do is to constantly give encouraging feedback no matter what they do, to push their confidence and let them "be" what they truly are. And, depending on the situation, trying to analyze the kids' actions without constantly staring at them and making them feel under pressure.

R: I know that you recently did some remote user testing with kids (due to COVID-19 social distancing), which is quite unusual. What are the biggest differences compared to testing in person? How are kids responding to this kind of testing environment? Any tips to share on this?

M: Remote testing was something I didn't believe was going to work. I am happy to say I was very wrong. It worked very well and it even opened up a new testing perspective. The kids were always at home with their parents and extremely confident and relaxed. All the effort to make them "feel like home" wasn't necessary, and we were able to see them play in the environment where they usually play! In some cases, parents were too involved in their kids' testing, but after a few "virtual playdates" we became experts in managing both parents and kids!

It worked so well that we decided to keep doing remote testing in the future. However, it is important to underline that this is not a substitute for in-person testing but it is an effective add-on.

Chapter Recap

- User testing is different with kids and adults.

- Even a small age gap can be significant when doing user research with children.

- Recruit more participants than the ones you think you'll need.

- Try to have a variety in personalities, introverts and extroverts.

- Parent's consent is mandatory.

- Set up a welcoming child-centered environment, either in a school or public library or a testing facility.

- Make the environment friendly, but not too distracting.

- Dress professionally, but casually. Overdressing might be intimidating.

- Try to include some popular characters in your outfit. It works as an icebreaker.

- Get the trust of the parents. Explain what the test is about (which is not about their child) and how it will work.

- For younger kids (2 to 4 years old), the parent or guardian could be in the room if this put the child at ease.

- The parent in the room should be a silent observer, away from the participant's line of sight.

- Ask for parents to intervene when support and encouragement is needed, but pay attention they don't influence the kid's choices.

- Some kids benefit from working in pairs with a close friend.

- When working in pairs, establish the turns so that there are no fights over using the device.

- High-fidelity prototypes are the best options with children.

- Children tend to explore and tap on everything. Include a quick way to reset the prototype.

- Let participants know they are not being tested, the product is. They are the teachers.

- Start with easy questions and tasks to build confidence in the tester.

- Keep questions concrete and use an age-appropriate language. Don't use technical or difficult words.

- Let them know it's OK to not know an answer or how to complete a task.

- No one's feelings will be hurt by negative comments.

- Allow for a little exploration outside the task boundaries (if the prototype allows it).

- Sessions should last a maximum of one hour.

- Account for a longer time for user research than when testing with adults.

- Plan for a toilet break right at the beginning and include one or two breaks during the session.

- Explain what the test will be like.

- Provide generic but positive feedback. No "poker face," like in testing with adults.

- If kids have a hard time expressing their feelings, try to use a visual scale with emoji-like pictograms or illustrations.

- If possible, focus on the participant only, while someone else takes care of annotating.

- Video recording sessions could be helpful to review answers and body language.

Market Your Product

An Overview on Strategies to Monetize and Sell.

Of course, I want to sell this record—there's no point making it otherwise.

—George Michael, British musician

Let me say right off the bat that this is not a business strategy book, so what we'll be focusing on in this chapter will be those business decisions that have a direct impact on the design of the product. Sure, we'll also take a look at some tips and marketing strategies, but for more in-depth information about the business of digital products, there are a gazillion of other books and authors.

First of all, the monetization strategy of your product. This is not something you can decide later on, once the product is finished and ready to be launched. Yes, you'll still be able to pivot later, facing all the redesign that will be necessary, but it's something you need to have in mind since the very beginning when launching the product for the very first time. Knowing whether it will be a free app, a paid app, or free + paid premium content, or based on subscription, and so on has an impact on how you design the experience. We'll take a look at what these models mean, how they work, how your design should be shaped around them.

© Rubens Cantuni 2020
R. Cantuni, *Designing Digital Products for Kids*,
https://doi.org/10.1007/978-1-4842-6287-0_10

Another topic will be the one about certifications. Certifying your app as safe and suitable for children's use is not just a way to check and validate your ethical UX decisions but also an important marketing tool toward parents and educators (yea, kids won't care so much about that).

We'll discuss about engaging educators and get their help by becoming ambassadors, and we'll take a look at some other tips and things to keep an eye on to help our product get some traction.

Business Models

There was a time when software products were physical boxes on shelves. You picked them up, handed over your money at a register, and that was pretty much it. It wasn't much different than doing groceries at your local supermarket.

Then things evolved (and I'm using this term carefully) into many more options for the consumers and strategies for the developers, especially with the advent of the Internet. *Shareware, freeware, adware, crippleware, trialware, donationware*… Lots of "wares," each one presenting pros and cons, and depending on the business strategy, the target audience, and the product itself to decide which way to go.

Digital distribution platforms, such as Apple's App Store or Google Play Store, Steam, and so on, introduced even new ways of monetizing digital products. Getting rid of physical packaging and physical stores made products easier to update and opened up to new opportunities such as *in-app purchases*. Nowadays, all products fall into one of the monetization strategies we'll analyze in this section, and we'll take a look at the reasons to pick one approach rather than another.

Paid Apps

This is the most obvious option, consisting in a price paid upfront upon download. It's the digital translation of picking up the box from the shelf and paying at the register.

While this is indeed the most straightforward monetization strategy, you need to consider that paid apps represent a very small minority of all the apps out there. Even big companies, like Adobe, in recent years moved to a different model (subscription).

The thing with paid apps is that their value for the customer must be absolutely evident before the purchase, and this is not easy to get, especially for a brand new product or, even worse, a new developer without a successful track record of other products.

For each product, there are surely countless competitors adopting an approach where customers don't have to pay upfront (free, freemium, in-app purchases, etc.). For this reason, if you opt for the paid app model, the quality of your product must be clear, it needs to stand out of the competition and be so much better to justify people to open their wallet without trying it first.

Toca Boca is once again a good example. The quality of their products is so high, their popularity, brand perception, and reviews are so stellar, that parents buy their apps paying upfront, sure to get what they paid for and probably more.

The biggest advantage of paid apps is the immediate return in terms of revenues, but the flip side is that getting downloads is way harder.

When to use

- You have a strong, authoritative brand.

- You clearly offer a better value over competitors, especially over free alternatives.

- You can afford a big marketing and advertising investment.

When to avoid

- Your product is content centered. You update content constantly.

- You plan to use other monetization strategies as well. Users that pay upfront don't expect any additional paywall or IAP (in-app purchase) within the product.

Freemium Apps

You might wonder why anyone would give away something for free. Well, there are several cases in which this could make sense. Considering that more than 90% of iOS and Android apps are free/freemium,[1] competing against this model is hard.

Why is this monetization strategy so popular? Free apps are more likely to be installed and tried out compared to paid app. This way, developers can lure in users with a free product that shows its potential, without delivering the full experience. In kids' products, this is a very common practice; the app is free, but children can play only one sample for each activity.

---

[1]Clement, J. "Distribution of Free and Paid Android Apps 2020." *Statista*. N.p., 15 June 2020. Web. 28 June 2020. <https://www.statista.com/statistics/266211/distribution-of-free-and-paid-android-apps/>.

Another example of freemium strategy is when an app is free but includes lots of advertising. The user can pay a price to unlock the *premium version* to get rid of ads. This is not common in products for children, because advertising is a very tricky (and unethical, in my opinion) way to monetize children's products.

Another strategy of freemium products is limiting the daily usage. *Lingokids* uses this strategy; children can enjoy three activities per day or get a subscription (more about this model later) to have unlimited access.

The power of the "free" price tag is very strong, and it can really help in getting the downloads. But once you get them, you need to convert, and here's the tricky part of the freemium model. The key is finding the balance between showing the value of your product and not giving away so much that users don't feel the need of moving to a premium version or a subscription. If you get tons of downloads and you struggle to get revenues out of them, it could be either that you don't show enough of your product to entice the users (assuming the product is good, of course) or that you give so much for free that they don't feel the need to pay to get even more. There isn't a magic formula to get this right, it's a process that involves market and user research, some trial and error, some A/B testing to find the right tune.

When to use

- You're a new developer, without enough credibility to go with a paid app.

- You want to combine multiple revenue streams (in-app purchases, subscription, ads…).

- You don't have a big budget for promotion, so you need lots of organic downloads.[2]

When to avoid

- Your brand or product is strong enough to get people to pay upfront.

---

[2]By "organic" we mean happening naturally, without the boost of ads, promotional posts, paid articles, or even black hat practices such as paid downloads and reviews (these are dangerous practices that could cause your product or account to be suspended from the distribution platforms).

Subscription

The subscription model is more and more popular nowadays. Many companies, big and small, rely on this revenue model in all kinds of industries. Think about Netflix, Spotify, Amazon Prime, Apple Arcade, Adobe Creative Cloud, we're surrounded by subscriptions.

This strategy is good for developers as subscriptions provide a rather stable flow of revenues, but what's the advantage for the user? This has to be clear, people don't want to pay every week, month, or year if they don't perceive a clear advantage in doing so. Not all products are suitable for this monetization strategy; there must be some kind of constant update to the product. Therefore, this option is good for products that have content at the center of their experience, content that is constantly updated. *StoryBots* worked on a subscription, because our content library was regularly refreshed with new videos, new books, and more. Another reason to justify a subscription is a steady flow of new updates in terms of features or services; this is the case, for example, of Adobe Creative Cloud.

There are two classic approaches to promote a subscription:

- The freemium approach described earlier, in which users can try a limited version of the product and see part of the content locked until they subscribe (e.g., *Lingokids*).

- The free trial approach. In this case, users get full access for a limited period of time (e.g., *Papumba*).

The difference with these is that with the freemium approach, users will continuously access your app over time and are constantly reminded of its value and what they would be getting with a premium account. With the trial period strategy on the other hand, users will be forced to create an account, and this could give you a lead to push marketing activities. Sure, you could combine the two and require an account also for a freemium version, but adding a mandatory registration to access a limited version of the app could be asking too much to the users and inhibit their interest into testing the product altogether.

The popularity of this model though is now getting to a point where people are thinking very carefully to subscribe even a $2.99/month service, as this is going to add up to many other subscriptions with the feeling of losing control of them. There are several products today that make a selling point of not being based on subscription, a onetime payment is becoming more and more a strong advantage in the eye of the user.

When to use

- The product is continuously updated with new content and/or features.

When to avoid

- Your app is not content centered.

In-app Purchases

In-app purchases, or IAP, very often work together with the freemium model. Unlocking the full app (its "premium" version) is usually done via an IAP. But IAP can be used to deliver all sort of digital goods. Besides unlocking the full version, you can sell individual packages for a smaller price tag. By providing more options, the user can customize the experience cherry-picking only the features and the content they need. Normally you want to make the premium unlock more inviting as that'll be the most expensive item in the menu. So, let's say you sell three packages for $2.99 each, you should sell the premium version to unlock them all for something like $6.99, a classic marketing strategy.

When we talk about in-app purchases, we talk about real transactions, with real money, not to be confused with pretend purchases done with in-game currency, using "points" (or gems, or hearts, or stars…) kids earn by playing. IAP can't be performed by children; all the shopping activities must be done by an adult behind the parental gate (see Chapter 6). This adds a lot of friction compared to using this monetization strategy with adults, because we can't talk directly to the user (the child), but we need to find ways to get the parent doing the purchase. Some apps add a voice-over inviting the kid to ask the parents to buy the item for them whenever they tap on a locked feature or piece of content.

One thing to consider is that, just like it happens with paid apps, the distribution platforms (App Store, Google Play Store, etc.) take a commission (usually 30%) on each IAP. So keep this in mind when putting a price tag on your items. The good news is that, according to *Sensor Tower* (the leading platform in analytics for the business of apps), users are spending more and more in purchases inside apps, a market that grew by 17% in 2019, compared to the previous year.[3]

---

[3]Nelson, Randy. "Consumer Spending In Mobile Apps Grew 17% in 2019 to Exceed $83 Billion Globally." *Sensor Tower Blog*. N.p., 06 Jan. 2020. Web. 01 July 2020.<https://sensortower.com/blog/app-revenue-and-downloads-2019>.

When to use

- In-app purchases are a perfect complement to the freemium model.
- You have a catalogue of digital goods that people may want to buy to enhance their experience.

When to avoid

- If you decide to go with the paid app model, users expect to have all the features without additional purchases.

Advertising

Despite advertising being one of the major revenue drivers in digital products business, it's not an easy way to monetize products for kids. I talked extensively about this in Chapter 6 when discussing the safety measures in children's products; I just feel I should mention this option here to have a complete overview of the most common monetization strategies on digital products.

Yes, there are networks specialized in delivering ads suitable for kids, but there also are strict policies on App Store and Google Play Store, about advertising to children. I personally make this a matter of ethics, most of all. Don't forget that children don't have our grown-up experience in discerning what is an advertised content and what is a genuine piece of content of your product. This may lead safety gates popping up unexpectedly and confusion in general on the app usage, not to mention parents are not big fans of ads.

Certifications and Standards

When you sell a digital product for children, you speak mainly to parents and educators, as they are the ones making the purchase or suggesting it. Sure, you can entice kids with nice pictures, engaging videos, and pretty colors, but ultimately it'll be an adult deciding to download your product or not.

Parents can be more or less informed about what's suitable for their child; some parents read reviews, articles, talk to educators to make cautious choices. Others take a less informed approach and just want something to keep their kids busy for some time. As product designers, we should help parents and make products that are safe and ethical. While the first kind of parents wants to be reassured about your product's safety, the others need to be informed and educated on what to look for in a digital product for kids.

Certifications can surely help you assess the level of safety of your product, but they also serve as a marketing tool toward parents and educators, to let them know that your product is indeed suitable for kids and certain measures are in place. Let's take a look at some of the most popular and trusted ones.

kidSAFE

kidSAFE maybe the most well-known seal for children's products. Founded by Shai Samet, a children's privacy consultant, kidSAFE offers a series of seals to certify compliance to different kinds of safety measures in kids' digital products. There are currently three seals.

kidSAFE Certified Seal

This requires compliance with the following "Basic Safety Rules" (as applicable):

- Safety measures for chat, community, and social features
- Rules and educational info about online safety
- Procedures for handling safety issues and complaints
- Parental controls over child's account
- Age-appropriate content, advertising, and marketing

kidSAFE + COPPA Certified Seal

This requires compliance with the Basic Safety Rules, plus the following additional "COPPA Privacy Rules" (as applicable):

- Neutral age questions
- Parental notice and consent procedures
- Parental access to child's personal information
- Data integrity and security procedures
- COPPA-compliant privacy policy
- COPPA oversight and enforcement by the kidSAFE Seal Program

kidSAFE Listed Seal

This requires compliance with the following general principle:

- Designed and intended for use by children, families, and/ or schools

TRUSTe Kids Privacy

TRUSTe Kids Privacy is a service by TrustArc, which has been releasing TRUSTe certifications for websites for decades now (since 1997). Their TRUSTe Kids Privacy is, as the name clearly indicates, the children's version of their classic seal. They offer full assessment on COPPA compliancy for websites and apps targeting children under 13, assistance on remediation in case any changes are needed, monitoring, and guidance.

Privo

Privo is a company providing personalized consulting in the field of compliance and certification serving some of the biggest players in the kids' digital products industry. They offer a series of seals, including one specifically dedicated to GDPR, unlike kidSAFE offering only COPPA certification. Interestingly, Privo also offers API and SDK solutions so that developers don't need to worry about being constantly up to date with legislation in different countries.

ESRB

If you're a gamer, you know this very well. The Entertainment Software Rating Board (ESRB) is the association providing the seals you see on basically any video game on the market, because, despite being voluntary, all console manufacturers and some US retail stores and digital stores require this seal on each and every title.

There are five seals:

- "E" (Everyone): The content is for everyone. There might be a very mild cartoonish kind of violence (think classic Hanna-Barbera style of slapstick).

- "E10+" (Everyone 10+): For everyone older than 10. The level of violence in actions and language is minimally higher compared to "E."

- "T" (Teen): For 13 years old and up. Include higher level of violence, minimal presence of blood, simulated gambling, suggestive themes, and crude humor.

- "M" (Mature): The content is for 17 years old and up. Intense violence, blood and gore, sexual content, and strong language.

- "A" (Adults Only): The product is suitable for 18+. Basically everything is allowed, from prolonged violence to graphic sexual content and gambling with real money.

The ESRB is the seal many parents ignore (to go back to the introduction of this section) when they buy *Grand Theft Auto V* for their 13-year-old child, for example.

There are several reasons why you may want to get one (or more) of these certifications. First of all legal ones; you can of course do everything on your own and comply with all laws and regulations, but having a professional assessing the status of your product is for sure a safer option. Secondly, the marketing opportunity they offer. It's true that many parents are not really aware of these standards and too often overlook the presence of these seals or they simply don't understand their meaning. But I believe things are slowly changing, online privacy is a very hot topic today, users of all ages are increasingly aware of the dangers and the best practices to protect themselves (and their loved ones). With the rise of privacy concerns along with the growth of the kids' digital products industry, I think these certifications will be more and more sought-after by developers and by parents and educators as well. Another important thing to consider is that while parents can't be sued for letting their kids use an app that doesn't comply with COPPA or GDPR regulations, schools and educational institutions might be, so they are absolutely interested in knowing if a product is compliant and safe to use or not. If you plan to sell your product to schools, you can leverage on these certifications more effectively than you would with parents.

Teachers As Ambassadors

In the beginning of this book, we've seen how educators are one of our three major audiences, along with kids and parents/caregivers. While they might not have the same spending power of some parents (educational institutions, especially public ones, are often working on a tight budget), they can play a role as critical allies for us.

For example, it's a very common practice, among educators, to share tools that they've found helpful and they regularly use in their day-to-day, in and outside the classroom. Word of mouth in the teachers' community can be a powerful aid to your marketing strategy, and by "word of mouth" I also mean online, on dedicated forums, newsletters, and blogs. Some teachers have personal blogs to share this kind of tips with their peers, for example, freetech4teachers.com is a blog by a teacher sharing ideas to leverage technology in the classroom with free resources.

"Free" is important in this context. Giving free access to your product to educators is a classic strategy, and it serves two purposes:

- They can help you by testing your product, from the first MVP[4] and forward, flagging bugs, giving feedback and suggestions.

- If they appreciate the product, they can spread the word, not just to other teachers but also to parents to try it at home.

Educators are an authoritative voice toward parents; they have their trust and (usually) follow their suggestions. For this reason, it's important that we gain educators' trust in the first place. How do we do that? Well, first of all with good products. Teachers are happy to jump on board and "sell" your app if it's a good product, with real value for the kids. We can also get their trust by involving them early in the development process. I can't stress enough how important it is to have an educational expert in our team, regardless of the product being explicitly educational or not.

Have a selected number of teachers joining your beta-testing pool and work along with them, listen to them, and learn from them. It's a situation where you can only win. If your product is bad, you'll know it earlier and you'll have better chances of steering it to the right path. If it's good, you'll get educators excited to share it and suggest its use to parents and other teachers.

Partnerships with schools and teachers can also be useful when you need to do user testing with children. Schools you gave free access to your product might help providing an environment for conducting tests as well as put you in touch with parents for recruiting children to test with.

Get educators involved requires more than just giving them free access to your app or website. You need to engage them and create a community around them. They need to know how to get in touch to report bugs and tell you their experience with the product, share opinions and tips with other educators, be up to date with new features and improvement, request new features, and so on.

Lastly, providing schools and teachers with free tools for their job can also provide a ROI in terms of brand reputation. Big tech companies like Apple and Google are giving away tools, platforms, and devices to educational institutions. I'm not saying these are just marketing stunts, there are for sure ethical decisions behind, the good old idea of giving back to the community, but it's undeniable that they can find a marketing use for these good deeds.

---

[4]Minimum viable product.

More Marketing Tips

As I mentioned in the introduction of this chapter, this is not a business strategy or marketing book (and being toward the end of it, you should know already), but it was important to touch some topics, like the monetization strategy as these influence the way you design the experience.

There is a lot more to say about the marketing of children's apps, but that could fill another book or even more than one. Plus I'm a maker, not a seller, so you will easily find someone more qualified than me to talk about such things.

Here I'll just point out a few tips and tricks to get a taste of it, in case you'll want to dive deeper into marketing matters.

Intellectual Property

Creating a brand, including characters and storytelling, from the ground up and making it a hit is a big challenge, especially if the first touchpoint is a digital product. Some video games (not specifically made for kids) did that, think of *Angry Birds*, *Pokémon*, or even the *Tomb Raider* franchise, just to name a few. They started as video games (a.k.a. digital products), but they became a pop culture phenomenon, with feature films, merchandise, and a brand worth billions. Children's digital products are far from that kind of success (yet). Toca Boca, with the series *Toca Life*, started venturing outside the boundaries of the screen with some merchandise, but as a brand they are still very far from any Nickelodeon's or Cartoon Network's intellectual property, not to mention Disney's.

The other way around though is an "easy" (if you get a partnership) way to boost downloads. *Peppa Pig*'s series of apps are OK products, but honestly far from being groundbreaking experiences; nevertheless, they're easily at the top of the App Store and Google Play Store charts in the kids' category. To be brutally clear, your greatly crafted product with your own characters will never be as successful as a mediocre product featuring Elsa from *Frozen*. Not even close.

So, if you manage to strike a deal with the owner of one successful TV show or toys franchise or movie, it'll boost your downloads greatly. The flip side is that IP owners don't give these away cheap; normally they commission the product to a developer for a fixed fee or for a small revenue share. So more downloads for sure, but it's not said it'll be more money.

For more insights on IP partnerships and kids' digital products, read the interview with Catriona Wallis later in this chapter.

Specialized Blogs and Publications

A lot of parents today have blogs on parenting and they often cover technology as a topic. The same goes with educators. These blogs talk to very specific niches and readers are there to look on tips about new products for their kids. Getting a review on these kinds of publications is sometimes more effective than being covered by high profile and more generalist blogs.

Parenting Groups on Social Networks

Similar to blogs, there are tons of groups on social media dedicated to parenting and education you can look into. Again, people who follow those groups are there specifically to get suggestions also on tools and new technologies.

Awards

Awards can be a badge of honor you can display on your website, App Store and Google Play Store product pages, social media pages as well as printed material, and more. If a product won one or more awards, there must be something good about it, right? That's more or less what you want prospect users to think. There are many awards out there, some more specific for kids' products, but maybe less popular, and some others more generic, but more popular to the general public.

The most famous awards for digital products, like the W3 Award and the Webby Award, have specific categories for children, while others, like the Parents' Choice Award or Teachers' Choice Award, are specifically dedicated to products for children and youths.

Industry Insight: Interview with Catriona Wallis

Catriona Wallis is founder and CEO of Colto. Being a mom with teaching experience, she noticed the potential of the iPad as an educational tool and founded Colto, a company that now delivered several award-winning titles for kids and signed partnerships with Nickelodeon and Highlights.

Rubens: I've been lucky enough to collaborate with Colto on several occasions, the first one at the very beginning of the company's history. How did the market for kids' digital products change in the past few years? Do you think apps got better lately in terms of quality, safety, educational value? Do you have more competition?

Catriona: As kids' app developers, we have a responsibility to strive to raise the bar on the quality of the products we put in the hands of children.

Before answering your question about recent changes, I wanted to note that the kids' app market is still considered early stage, given that apps were first launched in 2008 on the iOS App Store. There have been a lot of changes in 12 years, but there are still important issues that have not progressed and been resolved yet such as the difficulty with discoverability of good quality apps and the lack of regulation of apps by an independent third party.

The most notable change in the last 4 years is the monetization model (or revenue model) of kids' apps. In 2016, Apple introduced a subscription model for kids' apps. This rapidly became the most popular revenue model in the market, mostly because it was the first model that allowed developers to build a sustainable business with its recurring revenue. Prior to subscription, only transactional download revenue models were available (free, premium/paid, freemium, or hybrids of these). These single transaction models provided limited revenues for developers so it was unprofitable to acquire new users through paid acquisition which made it difficult for new developers to boost their apps' rankings to get the visibility for downloads and revenues. Before the introduction of subscription, we saw a lot of smaller independent developers enter the market and quickly disappear.

Another rapid change in the past few years linked to the introduction of subscription is that many more high-quality educational apps have been published in the kids' category. The sustainable nature of subscription led to greater investment by both larger and smaller developers in apps with an educational objective, particularly school readiness apps for preschoolers teaching pre-numeracy and pre-literacy skills. It should be noted here that the preschooler market is a particular one where the parent is the customer and the child is the user. The top-grossing list of kids' apps in the 5 years and under category is dominated by subscription model apps to learn the ABCs, 123s, as well as English language, speech therapy, and more. In the older kids age categories, children have more choice in the apps they download and you find more of a balance between educational and entertainment-based apps in the top-downloaded and top-grossing lists.

Two other important changes in the past few years are in the quality and safety of kids' apps.

The quality of kids' apps has definitely improved on iOS (less so on Google Play) due partly to the higher investment in apps since the introduction of subscription model and partly to Apple and Google becoming increasingly stricter in their app approval process.

The safety has fortunately improved in recent years with Apple introducing restrictions such as compulsory parent gates for kids' apps and age gates for apps with advertising as well as restrictions on data tracking for privacy reasons introduced in 2019 (although I worry that this just deters app developers from placing their apps in the kids' category).

R: Starting from its name, Colto, which means "cultured" in Italian, always had the educational value of its products at heart. Finding the right balance between fun and learning is the holy grail for children's apps. What's your advice to get this right?

C: When we create games with an educational objective, one of the greatest challenges is finding the right balance between making the gameplay fun and educational at the same time, as the two are often in conflict with each other. For example, if a game has too many educational assets, it can become distracting for the kids and lose their interest in the central gameplay. One experience we had which illustrates this clearly was during the development of Colto's app Highlights Monster Day, a game in which you adopt a monster for a day and care for it. The learning objective of the game is to teach prosocial skills such as empathy and kindness as well as healthy habits such as brushing teeth and eating healthy food. The game starts in the morning in a bedroom scene where you wake up your monster then scrolls right to different scenes as the day progresses. I wanted to have a clock on the wall on the UI of each scene which showed the different time of the day to help kids learn how to read the time. We already had the sun rising and falling during the day in the window of the background of each scene. My partner, our head game designer, didn't want to add the clock because she felt it was too distracting for preschoolers and compromised the simplicity of the user experience. We decided the gameplay experience was the priority in a game that teaches soft skills so we never added the clock.

R: In Colto you have the opportunity of experiencing two ways of making products: you develop your own, and you collaborate with famous IP owners like Nickelodeon and Highlights. What's the main difference in terms of creative freedom? Do these companies trust you completely, or are they giving more boundaries?

C: The amount of creative freedom varies a lot depending on the brand. In the case of Nickelodeon, our production team worked closely with their digital producer, and we had to follow the brand style guide for the two IPs we created apps for: Dora the Explorer and Nella the Princess Knight. So in the design we had very little creative freedom. However, in the game concept, we had almost complete freedom as Nickelodeon gave us no brief on what type of game they wanted so we created the game concept from scratch and pitched it to Nickelodeon then tweaked the concept together with their producer.

In the case of *Highlights for Children*, we had a lot more creative freedom in both the design and the game concept. With the first three apps, all the characters and assets were our original designs as well as the game concept. The fourth app we created with Highlights was a subscription app called Hidden Pictures Puzzle Play for their famous classic hidden object game. We had a lot less creative freedom on this product because Highlights had already built two preceding mobile versions of Hidden Pictures so they knew what they wanted for the product and gave us strict guidelines on the concept.

R: Talking about the success of a product, how much does having a popular intellectual property, like Dora the Explorer, help to make an app popular?

C: When we started working in partnership with famous brands, we heard from colleagues in the industry to expect a success rate of 10–20-fold for an app based on a famous IP such as Dora the Explorer. In practice, we have found the IP does give the app more visibility in the period immediately after launch which boosts the rankings and leads to more downloads within this 10–20 range. However, in recent years, there has been a massive increase in the number of apps based on popular IP, and the market today is flooded with branded apps making it more competitive. When we launched our Dora the Explorer app in 2016, you could get a boost just from the IP; nowadays, you need to have a generous marketing budget to acquire users even for apps based on popular IP.

R: Being the CEO of a company that makes apps for kids, and being a mom as well, what's your take on screen time? Do you set limits?

C: Much to my children's dismay, I do set screen time limits of 1 hour daily during the week and 2 hours on weekends. They say with parenting you have to pick your battles and this is one of mine. I do believe on the one hand that screen time is valuable for my children because it teaches them skills that are fundamental for the digital world they live in and keeps them in touch with pop culture. However, I've seen enough negative behavior from my children after prolonged periods of screen time, particularly on fully immersive games such as *Fortnite* on the Xbox, to convince me that it's not healthy for their brain development. I consider it important for my children to learn the discipline of having to switch off the device when their time is up even though they'd love to keep playing.

In my experience, it's easier to control screen time when children are 10 years and under in preschool, kindergarten, or primary school because most of their homework is still offline. When children are in secondary school, the lines become blurred because they use their devices for school work and it's difficult to control when they're playing and when they're actually working.

During the COVID-19 lockdown, all my usual screen time rules went out the door because the children's daily schooling was all online.

R: Which are, in order of importance, the best features to promote when marketing a kid's digital product to parents? Educational value? Safety? Entertainment? What else?

C: Definitely educational value remains the most important feature to promote, particularly to parents of children aged 5 years and under. Safety is a feature that I always believed would be more important to promote than I've found in practice is. Parents will tell you safety is important to them. For parents of younger children who are in control of which apps their children download and purchase, they want to know the app is safe. However, I think the reality for most parents of children aged 6–12 years is they don't have enough time in a day to read the App Store description and check the safety of every app their child downloads. I think it's always important to promote apps, particularly educational apps as entertaining to both parents and children, as they need to want to play the app in order to download it. If a child really wants an app because they've heard about it from a friend, they will nag their parents to have it until they do.

R: Last question: what's going to be the future for this industry? What should the focus be?

C: I'm glad you asked this question about what the focus should be because as mentioned in my first answer, while there have been changes in the kids' app market, there are some aspects that need to change in order for this market to progress and grow.

Firstly, there has never been regulation of kids' educational apps by a neutral third-party organization. This means that you have educational apps on the stores that claim, for example, to teach your child to read which have not been created by educational experts alongside educational apps that have been created by a panel of teachers and based on a school curriculum. In the absence of any rating or classification of educational value, it's difficult for parents to find good quality apps, and they might download an app from a developer claiming that it will teach their child to read, for example, and be disappointed by the results and incorrectly assume that all educational apps are low quality. Given that educational value is the most important factor for parents in deciding which apps to download and purchase, I believe there needs to be an independent assessment of the learning value that sets a realistic expectation of what results parents should expect from educational apps.

The major mobile platforms such as the App Store, Google Play, and Amazon App Store should focus on collaborating more closely with developers to incentivize them to place their kids' apps in the kids' category because new rules such as Apple's sudden announcement in June 2019 that all kids' apps can't include third-party analytics software (which are needed to track data that is vital to the app's success) deter developers who have created games for kids that can also be played by adults, from publishing their games in the kids' category.

Educational apps will continue to grow in number and popularity, given that educational value has consistently been the number one factor for parents deciding which apps to download for their children.

Finally, with the rise of kids' brands on YouTube due to the migration of children from traditional linear TV to watching videos on YouTube and streaming SVODs (subscription video on demand) such as Netflix, there are new brands from kids YouTubers such as Ryan's World, Kids Diana Show, and CKN Toys that are increasingly publishing kids' apps. There are of course still apps based on evergreen classic kids IP such as Barbie and Peppa Pig; however, we will see the new wave of YouTube brands as well as new IP from kids shows on Netflix, Amazon Prime, and so on dominating the kids' app market.

Industry Insight: Interview with Raffi Frensley

Raffi Frensley is Marketing Director at GoNoodle, a company creating digital products for kids that promote physical activity, subverting the idea of screen time as sedentary time. Raffi has not just years of experience in engaging educators and creating communities with them, but has also a past as an elementary school teacher himself, with years spent teaching technology literacy and computer skills to kids.

Rubens: I'm really happy to have this chat with you because you're not just Marketing Director at GoNoodle, but you have also a past as an elementary school teacher. What's your take on the current landscape of digital products for kids?

Raffi: It's a wild time for digital content! There is such saturation and it's both exciting and overwhelming. Students have infinite access so the trick becomes directing students to vetted and approved content. I think we're also seeing a shift from delivering content (watching a YouTube video) to having kids create authentic experiences within the product (practicing and recording a TikTok). I think we're creating a generation that creates almost as much as they consume.

RC: What are teachers looking for in kids' interactive experiences? And what about the parents?

RF: Experiences should explain themselves. The value should be inherent quickly. I've seen so many robust tools with multiple functions (all very well crafted) that lacked a clear call to action. Start with a hook (Virtually Tour Ancient Ruins!) and then build out the user experience. ClassDojo originally started with a very simple premise—if you're awesome in class, you get a point. When you're not awesome, you lose a point. It was very effective and incredibly simple for students, teachers, and parents to understand. They've since built out their model to include curriculum, and a suite of other teacher tools, but they started with a simple premise. Also teachers need interactive experiences to be free (or very inexpensive) and easy to implement in their classroom. There's a lot of red tape around school budgets, and the more you empower teachers to be the leaders, the easier it is to onboard new users.

RC: I like GoNoodle because it goes completely in the opposite direction of what is the common idea of a digital product, meaning something that makes kids lazy and physically inactive. Considering the current (at the time I'm writing) COVID-19 situation, with lockdowns enforced in many countries around the world, did you see a spike in downloads and usage? Do you think this moment will make people realize we need more digital products designed specifically for children's needs?

RF: Yes! Oh my goodness. Sometime around March 16, 2020, as soon as families shifted to safe-at-home, we became a household name in about a week. It was wild. We already had a great cohort of over 1 million teachers who used GoNoodle in their classrooms. But once those students were stuck at home, teachers (and often students themselves) would recommend GoNoodle to parents. Why? Because it was something fun from their classroom that they could easily replicate at home. And as soon as they had a dance party with Mom or Dad, the parents got it. Even though GoNoodle had been in this iteration for about 5 years, parents were just learning about us. That good energy got us a lot of attention from media outlets from mommy bloggers to a shoutout on Good Morning America. I do think we need more digital products designed for children. There's opportunity: for everything from classroom furniture to violins to pencils, we make modifications for students to make resources developmentally appropriate. Why wouldn't this apply to digital literacy and digital resources? YouTube might not be developmentally appropriate for a small child, but a scaled down list of videos could be.

RC: What's your advice about engaging with teachers? How can we let them know about our product and become eager to try it out?

RF: Show off the value! I love fun and color just as much as anyone else, but teachers need to understand the value. I spent weeks trying to get teachers to test out a now defunct math app for SMART boards. I did trainings and tutorials and got a few sign-ups. One day a teacher posted a 10-second video of a student using the app in action and that day we had more sign-ups than the entire first half of the year! Maybe it was how easy it was for the teacher to set up. Maybe it was how quickly the student understood the task. Maybe it was their giant grin after answering correctly. Whatever it was, the brief video clip could better encapsulate the value of the product much better than any PDF guide I could have made.

RC: Can gaining teachers' trust and have them as ambassadors be an effective marketing strategy? If so, in short, what's the best way to make this happen?

RF: Treat your most engaged users like royalty! Not only will they share your product but they will give you critical feedback to help understand your users. Start small with a group of ambassadors and have regular communication with them. The worst experience is to pump up your ambassadors only to leave them in the dark and in silence for months with nothing but a T-shirt with your logo. Let ambassadors know about new content, let them know what's coming down the pipeline, let them provide ideas. Be as transparent as you can with your ambassadors because that's a microcosm for your product's community as a whole.

RC: How important is it the use of social media to create a community around your product? And what's the best way to increase engagement from teachers? And parents? Do you speak to them differently than to teachers?

RF: This is tricky! Social media is powerful and the community will exist whether you create it or not, so be proactive and say it first. I think it's easier to work with teachers because they have a common connection: they are all educators trying to teach. Being empathetic and in tune to the real problems that teachers face (like the cutting PE (physical education) in elementary schools) authentically connects you to educators. The tone with teachers should always be supportive. Parents have more diversity but of course they are connected by their search for developmentally appropriate resources for their children. For parents, my recommendation is don't talk to the kid, talk to the parent. Talk to them like you would talk to your best friend, and be cognizant of the diversity of experiences. Make sure graphics include a range of family structures and a spectrum of colors to avoid isolating potential users.

Chapter Recap

- It's better to plan how to monetize your product early during the concept design phase, as this will impact the way you design the experience.
- There are different ways of monetizing a digital product, each one with pros and cons.
- Deciding which monetization strategy to use also depends on the nature of your product.

- Paid apps ensure immediate revenues, but are less likely to be downloaded, unless you have a very strong brand or your product has a very clear advantage on competitors.

- Freemium is the most popular choice along with subscription.

- Freemium allows users to test the product before deciding to buy the full version.

- With freemium you need to find the right balance between letting users understand the value and not giving away too much for free.

- In-app purchases work well along with freemium.

- In-app purchases allow users to customize their experience, buying only the content and features they're interested in.

- Subscription is very popular today and ensures developers a steady stream of revenues.

- The subscription model works only if your product is constantly updated with new content.

- Advertising is the main source of income for many digital products today, but for kids is more complicated.

- Distribution platforms have strict policies regarding advertising to kids.

- There are ethical implications when using advertising in children's products.

- Certifications can assess the safety of your product and work as a marketing tool toward parents and educators.

- Teachers can be allies in promoting the product to parents and other educators.

- Letting teachers use the product for free will make on boarding them easier.

- Educators can be your beta testers since the very beginning and provide useful insights with their professional perspective.

- Create a community around teachers and engage them.

- The use of a popular IP can greatly boost the downloads of your app.

- Look into specialized blogs and publications on parenting and education for reviews and promotional content.

- Promote your product in social media groups for parents and teachers.

- Awards can be used to make your product look more authoritative and entice users to try it.

CHAPTER 1

Beyond the Screen

An Exploration on Experiences That Break into the Real World.

Most of this book is about principles that will still be valid for a very long time (the human factor influencing touch-based interactions, color theory, typography, visual perception, and so on, won't change anytime soon), but it's inevitable that new technologies will bring new opportunities for designing new interactions and, therefore, we'll see the rise of new needs and best practices. It's exciting, as well as difficult, to imagine what will happen next, because, as British sci-fi writer Arthur C. Clarke stated in the last of his famous three adages (also known as Clarke's three laws), "any sufficiently advanced technology is indistinguishable from magic."

We live in a marvelous time, when new technologies arise on a daily basis. The evolution of technology is nothing like the incredibly slow process happening in nature and described by Charles Darwin. It's made by giant leaps forward, in a logarithmic fashion more similar to what's stated in Moore's law.[1]

---

[1]In 1965, Gordon Moore, engineer and businessman, cofounder of Intel, noted how the number of transistors in a dense integrated circuit doubled every 2 years.

© Rubens Cantuni 2020
R. Cantuni, *Designing Digital Products for Kids*,
https://doi.org/10.1007/978-1-4842-6287-0_11

For almost two hundred and fifty pages I've been talking, for the most part, about what is going on within the screen, as that's where the vast majority of digital products for kids live nowadays. But, before we say goodbye, I'd like to invite you to take a look at some ideas for products that go beyond that. These technologies and products are out there today, but pushing the boundaries of what's the norm.

Augmented Reality

As of mid-2020, augmented reality (AR) has yet to make a big break into our everyday lives. This technology has been around for quite some time now, but we seem to struggle to find a real application for mass consumer market. The most successful implementation of this technology has been the mobile game *Pokémon Go*, which took the world by storm in 2016, becoming an instant pop culture phenomenon.

One of the struggles AR is facing is, in my opinion, the lack of a proper device to make it work. The overlapping of virtual objects on the real world requires the users to look and move around while peeking through a screen kept in front of their eyes. It's not really comfortable, especially if you consider that a smartphone is lighter than a tablet but offers a very small "hole" through which we can see the AR world, while the bigger screen of a tablet offers a more comfortable window at the cost of a heavier device to hold.

There are a few devices that perfectly work with this technology, for example, imaging using it while driving, with information appearing on the windscreen of our car. This is something all major automotive OEMs (original equipment manufacturers, or car makers, in this case) are working on right now. But probably the best device to enjoy AR will be smart glasses, another device in the works for some time now.

Until these will become mass market products, AR will always struggle to bloom. But augmented reality encapsulates a "wow" factor, something that embodies that "magic" Clarke mentions in his law, and this makes it a valid component for digital products for children. Kids love to see and experience magic, they love to be wowed (me too, actually), and, as I mentioned multiple times, kids are in for the journey. So while resorting to AR in products for adults (task driven) may often look gimmicky, doing so in products for children (experience driven) could give our product an edge.

The downside with using AR in children's product is, again, the ergonomics of it, in particular with younger children. Pointing at the target the right way, holding the device while maybe having to interact with on-screen elements, especially when using a tablet, can pose some challenges and it's something to keep in mind.

Nonetheless, let's see some examples of interesting implementation of AR in products for children.

QuiverVision

QuiverVision is interesting because it perfectly mixes a new technology with an "old school" activity such as coloring books. Grown-ups can download and print PDF documents from a very extensive archive of illustrations, and kids can color them with their regular markers or crayons (Figure 11-1). These illustrations come to life in AR as 3D models by pointing at them with the device's camera through QuiverVision apps. There are a series of apps, like the classic QuiverVision, the educational one, the fashion designer one, and one for face masks.

Some of these illustrations, once brought into AR, even include interactions and small games, where the child can interact with the character and play with it.

The mix of digital and non-digital is a perfect way to create that magic effect I was mentioning earlier. The 3D models kids see in augmented reality are customized with the colors and decorations they created with their hands on a flat illustration on a piece of paper, plus the possibility of interacting with the model. It's a very amusing effect.

Figure 11-1. QuiverVision app bringing a character to life

Geo AR Games

Geo AR Games is a developer from New Zealand working on some interesting concepts involving AR and geolocation. Their app *Magical Park* invites children to discover the world outside by exploring a real environment enhanced and gamified by AR (Figure 11-2). The combination of augmented reality and geolocation is similar to *Pokémon Go*, but the area where *Magical Park* works is limited to a designated public park, so it's safer for children. Just like with *QuiverVision*, it's interesting how AR enables a combination of digital play and traditional play, in this case in particular also promoting physical activity.

Figure 11-2. A dinosaur themed experience developed by Geo AR Games

Math Ninja AR

Japanese developer Fantamstick used AR to teach math in an educational game called *Math Ninja AR* (Figure 11-3). While the use of augmented reality in this app is more traditional, compared to the previous examples, it's worth noticing how the use of this technology makes the experience more engaging and fun for the users. All of the activities in the game could have been done without AR, in a completely digital environment inside the screen, but they decided to leverage on augmented reality to make something different and more magical for kids, resulting in learning math in a more enjoyable and captivating way.

Figure 11-3. Math Ninja AR by Fantamstick

Real Toys + Digital Toys

Digital toys and apps are not necessarily in competition with traditional toys. In fact, toys today are becoming more and more smart, and there are amazing products on the market that perfectly combine the best of two worlds.

Unlike AR that creates a fictional interaction between the physical world and virtual elements, these products involve the use of a real tangible toy that can be controlled, customized, or programmed with the use of an app. This idea is often applied with the intent of teaching the basics of coding to kids (and we saw some examples in Chapter 4), but there are many other applications both for entertainment and education.

Osmo

Osmo is a brand of smart toys that combine the use of a tablet with sets of physical pieces. Children can use these pieces to interact with what happens on screen. Osmo sets include a specific stand for the tablet and a special cap to place over the camera to make it point on the surface where the tablet is standing (Figure 11-4).

The app uses the camera to read what is going on on the table and recognize the pieces the children are using. There are sets for coding, math, shapes, spelling, science, geography, and more.

Figure 11-4. An Osmo set to teach coding

PlayShifu

PlayShifu products are another take on the concept of real + digital toys. Similar to Osmo, there are several kits, from board games to math, chemistry, music, and more. What's interesting is that there are three completely different products to use along with a tablet.

Shifu Plugo is a stand connected to a magnetic mat where kids can place different sets to interact with the games on screen (Figure 11-5). They can place pieces, connect them together, move them around, and the camera of the device will read whatever is happening on the mat to translate the action to the video game.

Figure 11-5. A space game with numbers using the Plugo Count kit

Shifu Tacto is a completely different product (Figure 11-6). The tablet lays on the surface in a special stand where two players can play against each other. It's a new take on board games, mixing classic mechanics and props (like figurines and other components) with video games. The games are educational, teaching STEM subjects such as physics with reflection and refraction, and chemistry with molecules and compounds or more classic and strategic.

Figure 11-6. Shifu Tacto set

Shifu Orboot is a very interesting product that mixes together an old-school educational tool like the globe, a tablet, and AR (Figure 11-7). Kids can use the tablet to interact with the globe in augmented reality and access information about animals, cultures, food, inventions, monuments, and maps around the world. More globes are coming, for dinosaurs and Mars.

Figure 11-7. Discovering monuments around the world with Shifu Orboot

Pigzbe

Pigzbe is a digital piggy bank for kids (Figure 11-8). Financial products for children are not so widespread yet, but they're getting increasingly popular. *Pigzbe* aims to teach children 6 years old and up about the value of money and saving. Parents or caregiver can assign a task from an app on their phone and define a monetary reward. When the task is completed, children see the reward being delivered in the form of a tree, representing their savings, getting watered and growing. The product also features a physical device: a very nicely designed modern version of a piggy bank that delivers notifications whenever a new task is assigned and displays the savings amount. This way children can keep track of tasks and their money without the need of checking the screen of the tablet too often.

Figure 11-8. Pigzbe's parents' app, children's app, and digital piggy bank device

Chapter Recap

- As new technologies arise, we'll see more opportunities for new digital products and new experiences.

- Digital products can go beyond the screen, mixing digital and real.

- AR can help us get that "wow" factor our product needs.

- AR can be used to combine activities done in real life with interactions on a touchscreen.

- Digital toys are not in competition with real toys. They can enhance each other.

- By using its camera and other sensors, a device can understand what the kid is doing and respond accordingly.

- Tablets can be used to create a new and more advanced generation of board games.

- The physical component of the product can be a companion device, not necessarily a "toy."

Conclusion

Digital products are now a big part of everyone's everyday life. They changed the way we do things, the way we communicate, we travel, we work, we study, we do sports, we find a partner, and so much more. Trying to shield kids from digital products would be not just anachronistic but simply wrong. Like many other things in the world, these products can be used wisely or recklessly, they can be designed to benefit the user or not. As designers, parents, educators, and adults in general, it's our duty to be sure children benefit from responsible products, designed around them, for them, and with them.

If done right, digital products can be a precious aid during a child's development, along many other activities. The key, as always, is in the quality of these products and the balance in their use.

During my years as a designer, I had the chance to work on many different kinds of projects, but the ones I did for kids are among the best ones I remember and the ones I enjoyed the most. By designing for children, we can learn a lot about designing with empathy and so much of the knowledge we acquire with these projects can be used on products for adults as well. Designing for kids, who haven't fully developed their cognitive and motor skills yet, can teach us a lot about designing for people with disabilities, for example. They can teach us to design for inclusivity, which something we can't overlook anymore.

We have seen how digital products for children are different from the ones for grown-ups and it's no joke to design them. I started this book with the idea of creating a compendium for product designers who want to design for

R. Cantuni, *Designing Digital Products for Kids*,
https://doi.org/10.1007/978-1-4842-6287-0_12

kids, trying to put in words years of work and research (not just mine but by many designers and scholars who, I hope, I credited correctly). The aim was to make a manual, extensive enough to be the main resource to help readers come up with a decent product. Of course, each one of the topics we touched could be greatly expanded and elaborated on, but I hope this can become one of your go-to references and you open and reopen this book during your career. This is the book I needed when I designed my first digital product for children and, hopefully, will be the book you'll need when you'll design yours.

Index

A

Abstraction, 166–169

Active learning, 70, 71, 74

Age *vs.* device peculiarities
    desktop-based products, 111, 112
    touchscreen, 109, 110
    unintentional touches, 110

American Academy of Pediatrics (AAP), 138

Augmented reality (AR)
    Geo AR games, 242
    Math Ninja, 242, 243
    meaning, 240
    overlapping, 240
    QuiverVision, 241

B

Bishop, Chris, 199–201

Business opportunities, 1–4

C

Children *vs.* adults
    cognitive development, 100
    digital products, 100
    fundamental similarities and
        differences, 101
    meaning, 99
    physical and cognitive level, 100
    user experience (UX) principles, 101

Clarke's laws, 239

ClassDojo, 20, 36–38, 234

Cognitive development
    characteristics/developmental
        changes, 104
    cognitive load/prevent errors, 107, 108
    color palette, 107
    concrete operational stage, 104
    existing mental models, 106, 107
    goal activity, 105
    instructions, 105
    meaning, 102
    Piaget, Jean, 102
    preoperational phase, 103
    procreate palette panel, 106
    stages of, 103

Colors palette
    brightness/saturation comparison,
        153, 154
    emotional reaction, 152
    gender bias, 164–166
    harmonies, 154
        analogous palette, 158
        complementary colors, 154, 155
        monochromatic palette, 159
        split complementary, 156, 157
        squared and rectangular, 157, 158
        techniques, 159
        triadic palette, 155, 156
    kids *vs.* adults, 162–164
    misconception, 151
    saturated tones, 153
    tools online, 160, 161
    wireframes, 151

Conceptual skeuomorphism, 150

Printed in the United States
By Bookmasters